T0073848

"This is a remarkable book by an unassailable grand master of sound perception and auditory illusions. The text is very clear and very lively. Finally a book on sound perception has the sounds right on the pages! Point your phone, hear the sounds, it's that easy. Not only the sounds, but explanations from the author in her own voice. I settled in and felt like I was having a conversation with her. Deutsch is a keen and careful scholar, yet manages to make the pages incredibly entertaining. When one reads this book, one realizes that Prof. Deutsch didn't 'get lucky' when she discovered her well-known illusions. There is a program, guided by deep knowledge and intuition. She shares both with us in this wonderful book."

—ERIC J. HELLER, Abbott and James Lawrence Professor of Chemistry, and Professor of Physics, Harvard University, author of *Why You Hear What You Hear*

"In this delightful volume Diana Deutsch, a living legend in the field of music psychology, invites us into her laboratory. There, with the help of web-based audio files, we can listen in as she tricks our hearing into revealing some of the inner workings of the human auditory system. Dozens of these musical illusions help us to understand the complexity and marvelous sophistication of how we uncover patterns and meanings in the sounds that we hear."

—ROBERT O. GJERDINGEN, Professor of Music, Northwestern University, author of *Music in the Galant Style*

"Diana Deutsch is a true pioneer. In this finely written and yet seriously scientific book, she tells the story of how she discovered phantasms that to our ears are as miraculous as a Fata Morgana is to our eyes. Read and wonder!"

—STEFAN KLEIN, Professor of Critical Studies, University of the Arts, Berlin, author of *The Science of Happiness*

"It is a great pleasure to have Diana Deutsch's pioneering work on auditory illusions and her landmark explorations of the influence of language on music perception brought together in the summation of a stellar career that has profoundly influenced the field of music psychology and cognition. The underlying thread throughout the book is the extraordinary complexity of the auditory system and the wide range of individual differences among listeners."

—JONATHAN BERGER, Denning Family Provostial Professor in Music, Stanford University

"Diana Deutsch's pioneering work on auditory illusions opened up a crack through which music and speech perception could be understood in new ways. This engaging volume, laced with anecdotes and firsthand accounts, should pique anyone's curiosity about how the mind hears."

—ELIZABETH HELLMUTH MARGULIS, Professor, Princeton University

"Dr. Deutsch has been one of the world's leading researchers of the psychology of music for over four decades. This book is the culmination of her stellar career of intriguing observations gleaned from her innovative investigative techniques. Her contributions to the field are on par with Oliver Sacks, Roger Shepard, and Jean-Claude Risset. Dr. Deutsch's rigorous yet charming style makes *Musical Illusions and Phantom Words* equal parts illuminating and fun."

—MICHAEL A. LEVINE is a composer for film, television, records, and games, as well as a concert music and musical theater composer. He has won eight ASCAP awards and was a Governor of the Television Academy (Emmys). He scored the CBS drama *Cold Case*, wrote the Spider Pig choir arrangement for *The Simpsons Movie*, produced Lorde's version of "Everybody Wants to Rule the World," composed the world's first pedal steel guitar concerto, and the Kit Kat "Gimme A Break" jingle.

"The Yanny-Laurel meme and other audio illusions actually say quite a bit about the perception of music and speech and the organization of the human brain. Diana Deutsch, the world's foremost expert on these fascinating 'perceptual anomalies,' makes compelling arguments for a variety of issues, such as that music and speech originated from a protolanguage; that our past experience unconsciously affects what we hear; that music theory can now be put to experimental tests. She has shown that absolute pitch, once thought to be completely hereditary and extremely rare, is not at all unusual among musicians in China, where a tone language is spoken. Anyone who has been mesmerized by Necker cubes and Escher prints will find this book engrossing and entertaining—it is a mind-expanding, ear-opening tour de force."

—PHILIP YAM, Science Editor and former Online Managing Editor for *Scientific American* Magazine

"From her early pioneering work to the present day, Diana's fascinating work and observations on music have captured our imagination and inspired generations of researchers. In this remarkably accessible and deeply engaging book, she expounds upon some of her most intriguing work on the varieties of illusions that arise in music and language, and what they tell us about the mind. This is a world where distinct melodies are heard in the two ears, even though only one was presented, where musicians suddenly experience auditory hallucinations of their own music, and where speech is mysteriously transformed into song. Captivating and profound, Diana Deutsch's book will be delight not only to researchers, but to anyone who is curious about the human mind."

—WILLIAM FORDE THOMPSON, author of *Music, Thought and Feeling: Understanding the Psychology of Music*

MUSICAL ILLUSIONS AND PHANTOM WORDS

# Musical Illusions and Phantom Words

## HOW MUSIC AND SPEECH UNLOCK MYSTERIES OF THE BRAIN

*Diana Deutsch*

OXFORD
UNIVERSITY PRESS

# OXFORD
UNIVERSITY PRESS

Oxford University Press is a department of the University of Oxford. It furthers
the University's objective of excellence in research, scholarship, and education
by publishing worldwide. Oxford is a registered trade mark of Oxford University
Press in the UK and certain other countries.

Published in the United States of America by Oxford University Press
198 Madison Avenue, New York, NY 10016, United States of America.

Library of Congress Cataloging-in-Publication Data
Names: Deutsch, Diana, author.
Title: Musical illusions and phantom words : how music and speech
unlock mysteries of the brain / Diana Deutsch.
Description: New York, NY : Oxford University Press, [2019] | Includes bibliographical references.
Identifiers: LCCN 2018051786 | ISBN 9780190206833 (hardback) | ISBN 9780197672280 (paperback)
Subjects: LCSH: Music—Psychological aspects. | Musical perception.
Classification: LCC ML3830.D49 2019 | DDC 781.1/1—dc23
LC record available at https://lccn.loc.gov/2018051786

Paperback printed by Marquis, Canada

# Contents

Contents

# List of Modules

## INTRODUCTION

Mysterious Melody                    http://dianadeutsch.net/ch00ex01

## CHAPTER 2

Octave Illusion                      http://dianadeutsch.net/ch02ex01
Scale Illusion                       http://dianadeutsch.net/ch02ex02
Tchaikovsky Passage                  http://dianadeutsch.net/ch02ex03
Scale Illusion on Xylophones         http://dianadeutsch.net/ch02ex04
Chromatic Illusion                   http://dianadeutsch.net/ch02ex05
Cambiata Illusion                    http://dianadeutsch.net/ch02ex06
Glissando Illusion                   http://dianadeutsch.net/ch02ex07

## CHAPTER 3

Galloping Rhythm                     http://dianadeutsch.net/ch03ex01
Interleaved Melodies                 http://dianadeutsch.net/ch03ex02
Timing and Sequence Perception       http://dianadeutsch.net/ch03ex03
Passage from
   Beethoven's Spring Sonata         http://dianadeutsch.net/ch03ex04
Timbre Illusion                      http://dianadeutsch.net/ch03ex05
Continuity Illusion                  http://dianadeutsch.net/ch03ex06
Passage from
   Tárrega's *Recuerdos de la Alhambra*   http://dianadeutsch.net/ch03ex07

CHAPTER 4

CHAPTER 5

CHAPTER 6

CHAPTER 7

CHAPTER 10

# Acknowledgments

OVER THE YEARS I have had the good fortune of discussing the issues explored in this book with many people. It would be impossible to name them all without turning this into a roll call, but I should mention that I have drawn inspiration from several groups of colleagues.

Those whom I met frequently at meetings of the Acoustical Society of America and the Audio Engineering Society, in particular John Pierce, Johan Sundberg, William Hartmann, Manfred Schroeder, Ernst Terhardt, Adrian Houtsma, Arthur Benade, and Laurent Demany, discussed the topics in this book with me from an acoustical perspective.

A remarkable interdisciplinary group of scientists, musicians, and technologists attended a series of workshops on the Physical and Neuropsychological Foundations of Music, which was convened in the 1980s by the physicist Juan Roederer in Ossiach, Austria, and here we formed lifelong friendships, and initiated the exchange of many ideas. Among those who attended the workshops, and others whom I met at the time, were the musicians Leonard Meyer, Fred Lerdahl, Eugene Narmour, Robert Gjerdingen, and David Butler; the physicists and engineers Reinier Plomp and Leon Van Noorden; and the psychologists David Wessel, Stephen McAdams, John Sloboda, Albert Bregman, Jamshed Bharucha, Richard Warren, W. Jay Dowling, Carol Krumhansl, Isabelle Peretz, William Forde Thompson, and W. Dixon Ward.

Later, together with Edward Carterette and Kengo Ohgushi, I founded the biennial series, the International Conference on Music Perception and Cognition. The first of these conferences was held in Tokyo in 1999, and it was there that I first met my friends and colleagues Yoshitaka Nakajima and Ken-ichi Miyazaki.

At that time I also had the pleasure of many discussions with Francis Crick, who was turning his attention to the study of perception. His office was at the Salk Institute, across from me at the University of California, San Diego (UCSD). He was fascinated by illusions, and on several occasions brought visitors over to my lab to listen to mine. I also had many fruitful discussions with the neurologist Norman Geschwind, who was especially interested in exploring the neurological underpinnings of the octave illusion.

Further, over the years, I have had many conversations with my colleagues at UCSD, notably Vilanayur Ramachandran, Donald MacLeod, Stuart Anstis, Gail Heyman, and Vladimir Konecni in the Psychology Department; Sarah Creel in the Cognitive Science Department; John Feroe in the Mathematics Department; and Robert Erickson, Richard Moore, Miller Puckette, Mark Dolson, Lee Ray, Richard Boulanger, and Trevor Henthorn in the Music Department. Other colleagues and friends with whom I discussed issues addressed in the book include Joe Banks, Xiaonuo Li, Elizabeth Marvin, Nicolas Wade, William Wang, and Isabelle Peretz. My graduate students have been a delight to work with, particularly Kevin Dooley, Jing Shen, Adam Tierney, Kamil Hamaoui, Frank Ragozzine, Rachael Lapidis, and Miren Edelstein.

I am grateful to a number of people for valuable discussions of issues that are explored in specific chapters. These include Scott Kim, Steven Pinker, Robert Berwick, Steven Budianski, Peter Burkholder, Andrew Oxenham, Jonathan Berger, David Huron, Gordon Bower, Psyche Loui, Daniel Levitin, Lisa Margulis, and Chris Maslanka. Adam Fisher, David Fisher, Joshua Deutsch, Melinda Deutsch, Temma Ehrenfeld, and Brian Compton read various chapters and made wise and helpful comments. Six anonymous reviewers generously gave their time to read all the chapters in detail, and provided important feedback.

I have benefitted greatly from my conversations with Jad Abumrad, the founder and cohost of the public radio program *Radiolab*. His farsighted observations led me to consider the broad implications of my illusions in detail. In addition, my extensive correspondence with the composer Michael A. Levine has been an especially valuable source of information, and his perceptive insights have helped me considerably in shaping the book for a musical audience. I have also had extensive correspondence with physicist Eric J. Heller, whose insightful comments and discussions concerning auditory illusions have been exceptionally valuable.

Other science writers with whom I have had very useful discussions include Stefan Klein, Ben Stein, Ingrid Wickelgren, Philip Yam, Philip Ball, Howard Burton, Charles Q. Choi, Shawn Carlson, Michael Shermer, and Oliver Sacks.

I am extremely grateful to Joan Bossert, my editor at Oxford University Press, for her wise guidance and encouragement; her detailed advice contributed substantially to the final manuscript. Phil Velinov and Keith Faivre at OUP have also been most helpful in shaping the final product. I am exceedingly grateful to Trevor Henthorn, with whom I have had many discussions concerning issues explored in the book, and who has worked expertly to create the modules containing the illusions of music and speech, together with recordings of my spoken commentary. I am also extremely grateful to Katie Spiller, who has provided considerable expert and detailed advice and help concerning many production issues.

Finally, I am grateful beyond words to my husband, J. Anthony Deutsch, whose support and feedback on my work over the years have been invaluable. It is to Tony's memory that I dedicate this book.

# About the Author

DIANA DEUTSCH IS a British American perceptual and cognitive psychologist, born in London, and is one of the world's leading researchers on the psychology of music. She is Professor of Psychology at the University of California, San Diego, and Adjunct Professor of Music at Stanford University. Deutsch is widely known for the illusions of music and speech that she has discovered; these include the *octave illusion*, the *scale illusion*, the *chromatic illusion*, the *glissando illusion*, the *cambiata illusion*, the *tritone paradox*, the *phantom words illusion*, the *mysterious melody illusion*, and the *speech-to-song illusion*, among others. She is also widely known for her work on absolute pitch, or perfect pitch, which she has shown to be far more prevalent among speakers of tone language. In addition, she studies the cognitive foundation of musical grammars, the ways in which people hold musical pitches in memory, and the ways in which people relate the sounds of music and speech to each other. Her many publications include articles in *Science*, *Nature*, and *Scientific American*. She is author of the book *The Psychology of Music* (1st edition, 1982; 2nd edition, 1999; 3rd edition, 2013), and of the compact discs *Musical Illusions and Paradoxes* (1995) and *Phantom Words and Other Curiosities* (2003).

Deutsch's work is often featured in newspapers and magazines worldwide; these include *Scientific American*, *New Scientist*, the *New York Times*, the *Washington Post*, *U.S. News and World Report*, the *Globe and Mail*, the *Guardian*, *Huffington Post*, the *Telegraph*, *National Geographic*, *Pour la Science* (France), *Die Zeit* (Germany), *Der*

*Spiegel* (Germany), *Die Welt* (Germany), *Forskning* (Norway), *NZZ am Sonntag* (Switzerland), and many others. She has given many public lectures, for example, at the Kennedy Center for Performing Arts in Washington, DC, The Exploratorium in San Francisco, the Fleet Science Center in San Diego, the Institut de Researche et Coordination Acousticque/Musique (Centre Georges Pompidou) in Paris, the Vienna Music Festival, the Festival of Two Worlds in Spoleto, Italy, the Royal Swedish Academy of Music in Stockholm, Sweden, the Strasbourg Festival of Art and Music, and the Haus der Kulturen der Welt, in Berlin, Germany. She is frequently interviewed on radio, such as by NPR (including *Radiolab*), NBC, BBC, CBC, ABC, German Public Radio, Radio France, Italian Public Radio (RAI), and Austrian Public Radio. She has appeared on television episodes of *NOVA, Redes* (Spain), and the Discovery Channel, among others. Her illusions have been exhibited in numerous museums and festivals, such as the Exploratorium, the Museum of Science (Boston), the Denver Museum of Nature & Science, the Franklin Institute (Philadelphia), the USA Science & Engineering Festival (Washington, DC), the Edinburgh International Science Festival, and other venues worldwide.

Among her many honors, Deutsch has been elected a Fellow of the American Association for the Advancement of Science, the Acoustical Society of America, the Audio Engineering Society, the Society of Experimental Psychologists, the Association for Psychological Science, the Psychonomic Society, and four divisions of the American Psychological Association: Division 1 (Society for General Psychology), Division 3 (Society for Experimental Psychology and Cognitive Science), Division 10 (Society for the Psychology of Aesthetics, Creativity and the Arts) and Division 21 (Applied Experimental and Engineering Psychology). She has served as Governor of the Audio Engineering Society, as Chair of the Section on Psychology of the American Association for the Advancement of Science, as President of Division 10 of the American Psychological Association, and as Chair of the Society of Experimental Psychologists. She was awarded the Rudolf Arnheim Award for "Outstanding Achievement in Psychology and the Arts" by the American Psychological Association, the Gustav Theodor Fechner Award for "Outstanding Contributions to Empirical Aesthetics" by the International Association of Empirical Aesthetics, the Science Writing Award for Professionals in Acoustics by the Acoustical Society of America, and the Gold Medal Award for "Lifelong Contributions to the Understanding of the Human Hearing Mechanism and the Science of Psychoacoustics" by the Audio Engineering Society.

MUSICAL ILLUSIONS AND PHANTOM WORDS

# Introduction

⌒

THIS BOOK IS about the auditory system—its remarkable capabilities, its quirkishness, and its surprising failures, particularly as revealed in our perception of music and speech. Over the last few years, scientists have made dramatic advances in understanding the nature of this system. These have involved research in many disciplines, including psychology, psychoacoustics, neuroscience, music theory, physics, engineering, computer science, and linguistics. I hope to convey to you some of the discoveries and conclusions that have resulted from these findings, from the involvement of different brain regions in analyzing the various characteristics of sound, to the principles by which we organize the patterns of sound we hear and represent them in memory. I will be focusing on my own research, and hope to communicate to you the excitement of carrying out explorations in this field, much of which is still uncharted territory, and where there are so many discoveries still to be made.

I will be writing particularly about the way we perceive and comprehend music, since it is here that the auditory system is at its most sophisticated. Imagine that you are in a concert hall listening to Beethoven's Fifth. Many instruments playing simultaneously are involved in creating the sound mixture that reaches you. How is it that you can hear the first violins playing one melodic line, the cellos another, and the flutes another? And how does your brain know which sounds to exclude from your experience of the music, such as coughs and whispers coming from the audience? These perceptual achievements must involve mechanisms of extraordinary ingenuity.

And consider what happens when you switch on your radio to your favorite station. The chances are that if a familiar piece of music is being played, you will recognize it immediately, even though you have cut in at some arbitrary point. To achieve such a feat of memory, your brain must have stored an enormous amount of information concerning many pieces of music—and must have done so with remarkable accuracy.

The central focus of the book concerns illusions of music and speech—their amazing characteristics, and what they reveal about sound perception in general. Illusions are often regarded as entertaining anomalies that shed little light on the normal process of perception. Yet I hope to convince you that the contrary is true. Just as the failure of a piece of equipment provides important clues to its successful operation, so perceptual anomalies—particularly illusions—provide us with important information about the system that generally enables us to perceive the world correctly.

Most people are familiar with optical illusions,[1] but they are much less familiar with illusions of sound perception. Yet as this book demonstrates, the auditory system is subject to remarkable illusions. Furthermore, people can differ strikingly in the way they hear even very simple musical patterns. These discrepancies do not reflect variations in musical ability or training. Even the most expert musicians, on listening to the stereo illusions described in Chapter 2, may disagree completely as to whether a high tone is being played to their right ear or to their left one. Here, disagreements often arise between right-handers and left-handers, indicating that they reflect variations in brain organization. Another illusion— the tritone paradox, which is described in Chapter 5—shows that the finest musicians can engage in long arguments as to whether a pattern of two tones is moving up or down in pitch. Perception of this illusion varies with the geographic region in which listeners grew up, and so with the languages or dialects to which they were exposed.

ᚙ

Why is hearing so prone to illusion? One surprising characteristic of the hearing mechanism is that it involves a remarkably small amount of neural tissue. Take the visual system for comparison. Patterns of light projected from the outside world give rise to images on the retina of each eye, and photoreceptors (rods and cones) in the retina send neural impulses to the higher centers of the brain. There are about 126 million photoreceptors in each eye, so the visual system can make use of an enormous amount of information that is present at the front end. In addition, a large amount of cortex—roughly one third—is involved in vision. Hearing is quite different. We have remarkably few auditory receptors (hair cells)—roughly 15 ½ thousand in each ear, with only 3 ½ thousand sending signals up to the brain. Compare this to the 126 million receptors in each eye! The amount of cortex devoted to hearing is difficult to determine, with estimates ranging from 3 percent to 20 percent, depending on which cortical areas are included in the calculation—however, it is definitely less than the amount of cortex that is taken up with vision.[2]

In addition, the brain has the formidable task of generating auditory images from the sound waves that reach our ears. The task is complicated by the nature of the environment through which sound must travel. If you are standing in a room, a sound that occurs in the room is reflected off the walls, ceiling, floor, and numerous objects, before it reaches your ears. Your auditory system needs to reconstruct the original sound at its source, taking into account the environment through which it has traveled. To complicate matters further, the patterns of sound waves that reach you differ depending on where you are positioned relative to the surrounding objects—yet you somehow recognize these diverse patterns as resulting from the same sound.

As a further issue, we are typically exposed to several sounds at the same time—for example, a friend speaking to us, a dog barking, a car traveling along the road, and so on. We therefore need to disentangle the component frequencies of the complex pattern of vibration that reaches our ears, so that we can identify each sound individually. The perceptual separation of simultaneous sounds presents us with a formidable challenge.

It is not surprising, therefore, that our hearing system involves a complex and tortuous set of neural pathways, which provide many opportunities to modify and refine the signals that impinge on us. This complex system enables an enormous amount of what psychologists call *unconscious inference* or *top-down processing*—a huge influence of experience, attention, expectation, emotion, and input from other sensory modalities—in determining what we hear.

One illusion that I discovered shows clearly the power of unconscious inference in hearing. A while back I proposed a neural network that recognizes simple pitch relationships, and I published an article suggesting that the brain employs such a network in music recognition.[3] After the article was published, I realized that my model makes a surprising and most improbable prediction; namely that we should be unable to recognize a familiar melody when it is played such that all the note names (C, D, F♯, and so on) are correct, but the tones are placed randomly in different octaves. This prediction seemed so counterintuitive that I concluded my model must be wrong. I went to the piano expecting to verify my mistake. I played a well-known tune with the individual notes placed in different octaves, and as

I had feared, I heard the melody without any difficulty. But by chance, a friend happened to walk past as I was playing.

"What's that you're playing?" he asked.

"Isn't it obvious?"

"It sounds like a complete jumble to me."

Then I realized what was going on. As predicted from my model, my friend was unable to form the musical connections necessary to recognize the melody. On the other hand, while I was playing the melody, I had an image in my mind of how it should sound, and by referring to this image, I was able to confirm that each tone was indeed correctly placed within its octave, so I perceived the melody without difficulty.[4] You can hear the octave-scrambled version and the correct one in the "Mysterious Melody" module.

Mysterious Melody

http://dianadeutsch.net/ch00ex01

This example illustrates that whether or not you are able to grasp a piece of music may depend heavily on the knowledge and expectations that you bring to your listening experience. There are many examples of this in vision. Fragmented figures, in which only some of the information is present, furnish particularly good illustrations. Look at the picture shown in Figure 0.1. At first it appears as a meaningless jumble of blobs. But as you go on looking (and this might take a few minutes) the picture gradually assumes a meaning, so that ultimately you wonder why you had a problem recognizing the image in the first place. And when you view the same picture again—even weeks or months later—the chances are that it will make sense to you immediately, since your visual system is organizing its various components in light of the knowledge that you had acquired.

Several chapters in the book demonstrate the involvement of unconscious inference in sound perception. In particular, the illusions explored in the early chapters are based in part on our perceiving musical patterns that are highly probable, even though they are heard incorrectly. Chapter 7 shows how illusions in speech processing are also due in large part to the involvement of unconscious inference.

Another theme, considered in detail in Chapter 3, involves the roles played by general principles of perceptual organization, first enunciated clearly by the Gestalt psychologists at the beginning of the twentieth century. These principles apply both to hearing and to vision. They include, among other factors, the tendency to form connections between elements that are close together, that are similar to each other, and that are related so as to form a continuous trajectory.

FIGURE 0.1. Ambiguous figure.

In addition, the book is concerned with how speech and music relate to each other, both in behavior and in brain organization. Two opposing views are considered. The first assumes that music and speech are the function of separate and independent modules, and the second assumes that they are the products of more general circuitry that subserves both these forms of communication. In this book—particularly in Chapters 10 and 11—I maintain that music and speech are each the product of modular architecture—indeed, that they are each the function of several different modules. In some cases, a module subserves a particular aspect of communication (such as semantics or syntax in speech, and tonality or timbre in music), while other modules subserve both speech and music (such as pitch, location, timing, and loudness). However, I also maintain that general high-level functions, such as memory and attention, are instrumental in determining whether a given sound pattern is heard as speech or as music.

Linked with this, in Chapters 10 and 11 I discuss the evolution of music and speech. Did they evolve independently or in parallel? Did one form of communication evolve from the other? Or did they branch off from a protolanguage that contained elements of both? This question preoccupied both Charles Darwin and

Herbert Spencer in the nineteenth century, and it has been debated ever since. As pointed out by Tecumseh Fitch[5] speech and music don't fossilize, and we don't have access to time machines, so we can only speculate about this issue. However, I discuss convincing arguments that speech and music were both derived from a form of communication—a musical protolanguage—that preceded ours.

A further issue is also addressed. Since the different attributes of a sound— pitch, loudness, timbre, location, and so on—are analyzed in independent modules, the question arises as to how the values of these different attributes are combined together so as to produce an integrated percept. This is one of the most fundamental and difficult questions concerning perception in general. So far, researchers have considered it primarily for vision—for example, in exploring how the color and shape of an object combine to produce an integrated percept—yet it is particularly important in hearing. In this book, particularly in Chapters 2, 7, and 9, I explore situations in which the attribute values of sounds are combined incorrectly, so that illusory conjunctions result.

The issue of the relationship between perception and reality runs through the book. As discussed earlier, the auditory system is particularly prone to illusion, and many illusions of sound perception are here explored. A remarkable characteristic of several of these illusions is that they are manifest by different listeners in entirely different ways. Some of these, such as the stereo illusions described in Chapter 2, vary in accordance with the listener's handedness, and so reflect variations in brain organization. The perception of another illusion—the tritone paradox, described in Chapter 5—instead varies with the language to which the listener has been most frequently exposed, particularly in childhood. So, taken together, these illusions reflect influences of both the listener's innate perceptual organization and also of his or her environment.

Individual differences in the perception of music are also manifest in extremes of musical ability. For example, as described in Chapter 6, absolute pitch occurs very infrequently in the Western world, even among professional musicians. Yet in China, where tone language is spoken, this ability is not unusual among musicians. Heredity also appears to influence the ability to acquire absolute pitch. At other extremes of ability, as described in Chapter 10, although most people perceive melody, harmony, rhythm, timbre, and respond emotionally to music, a few others are tone deaf, and here there is clearly a hereditary component.

The book also describes the inner workings of our musical minds that are independent of sounds in the outside world. Chapter 8 explores "stuck tunes" or "earworms"—the persistent looping of musical fragments as intrusive thoughts that arrive unbidden, and are often difficult to discard. Hallucinations of speech and music—which are even stronger manifestations of our inner mental life—are explored in Chapter 9. In addition to being intriguing perceptual curiosities, musical hallucinations shed important light on the way our hearing system functions in general.

〜

In addition to exploring the scientific findings that shed light on such issues, this book has another purpose. Throughout the ages, thinking about music has been driven by two very different approaches. On the one hand, music theorists have generally been system builders who laid down rules for composers to follow, without considering perceptual constraints. When justifications for musical rules were proposed, these tended to be numerological or mystical in nature.

In contrast, scientists have intermittently attempted to place music theory and practice on a firm empirical foundation, through carrying out experiments on how music is perceived and understood. Many of these earlier experiments are described in Eric Heller's comprehensive and engaging book *Why You Hear What You Hear*.[6] There was a particularly fruitful period around the turn of the seventeenth century, when scientists such as Galileo, Mersenne, Descartes, and Huygens made important discoveries concerning the perception of sound and music. But these attempts generally petered out after a generation or two, largely because of the difficulties involved in producing complex sound patterns with the precision necessary to arrive at firm conclusions. The sixth-century music theorist Boethius, a convinced numerologist, put the problem well when he wrote:

> For what need is there of speaking further considering the error of the senses when this same faculty of sensing is neither equal in all men, nor at all times equal within the same man? Therefore anyone vainly puts his trust in a changing judgment when he aspires to seek the truth.[7]

Even during much of the twentieth century, scientists investigating sound perception limited their research for the most part to the study of very simple sounds, such as pure tones heard in isolation, or possibly two at a time, or perhaps noise bands of various sorts. This approach made sense then, because it was only with such simple sounds that experimenters could be sure of obtaining reproducible results.

However, the revolution in computer technology over the last few decades has completely changed the picture, and we can now generate any combination of sounds that we wish, constrained only by the limitations of our imagination and ingenuity. So we are finally able to inquire into the ways in which we focus our attention when listening to sound patterns, the principles by which we recognize the sounds of different instruments such as a flute, a trumpet, or a violin, and the ways we analyze combinations of pitches in time so as to lead to perceptual equivalences and similarities. We can also inquire into the brain regions that are responsible for analyzing different characteristics of sound, using speech and music as signals.

Given these developments, I strongly believe, and hope to convince you also, that the time has come to place music theory on a firm experimental footing. When a music theorist promulgates a rule for composition, why should we accept the rule uncritically, or expect to recognize its truth by introspection alone? I argue that instead, the rule needs to be justified by demonstrating the advantages to the listener when it is followed, and the penalties that result when it is violated. I am not advocating that scientists should tell composers how they should compose. On the contrary, knowledge of perceptual and cognitive principles enable composers to make more informed compositional choices. So, for example, a composer could exploit an experimental finding showing that listeners hear a particular musical pattern as a random jumble of tones so as to create just such a perceptual jumble.

Put in a nutshell, I will be arguing that the only rational approach to music theory is to treat music as a creation of the human brain, and as perceived and understood by the human brain. And we cannot assume that the brain is a tabula rasa (blank slate) that will accept any type of musical information indiscriminately. As exemplified by the illusions I will be describing, our perception of music is influenced by severe constraints, which are imposed by the characteristics of the brain mechanisms that underlie these perceptions. Many of these characteristics are likely to be innate, while others result from exposure to our sound environment.

So when an orchestra performs a symphony, what is the "real" music? Is it in the mind of the composer when he first imagined the piece? Or in his mind after he has gone over it many times, and in the process changed his own perceptions? Is it in the mind of the conductor who has worked long hours to shape the orchestral performance, by having different instrument groups play separately as well as together, so that his perceptions have been altered during the rehearsals? Is it in the mind of a member of the audience who hears the piece for the first time? Or one who is already familiar with it? And how do we deal with perceptual differences between listeners, even on first hearing? The answer is surely that there is no one "real" version of the music, but many, each one shaped by the particular brain organization of the listener, and the knowledge and expectations that listeners bring to their experiences.

In this book, I explore a number of surprising characteristics of music and speech perception. I offer explanations for some of them, and indicate how they shed light on the workings of the auditory system. Yet there are other mysteries concerning hearing for which no sensible explanation has yet been advanced. For instance, we can't explain why most of us have stuck tunes in our heads much of the time. Perhaps there is no good explanation here, and the best we can do is assume that they serve as reminders of things that we need to do, or of things that we have experienced in the past. At all events, at a personal level it is exhilarating to explore unknown territory in hearing, to discover new phenomena, and to study their implications. If in this book I have managed to communicate some of the excitement that I experience in doing this, then I am content.

## LISTENING ALONG

As you read the book, it is important that you listen to each illusion when you read about it. The description of many of the illusions includes a module that gives you access to the appropriate audio file. The modules can be accessed via a smartphone or tablet using QR codes that are embedded in each chapter. You can also type the URL shown on the module to hear the sounds from your computer.

As described, some illusions are best heard in stereo through earphones or earbuds. Others are best heard through stereo loudspeakers. Yet others can be heard either through earphones or through loudspeakers. If you are using a laptop, smartphone, or tablet, it works much better to attach external loudspeakers or earphones to the device, rather than using the built-in speakers, because these may not be in stereo, and may not provide adequate sound quality. Set the amplitude so that the illusions are heard on the soft side—as a rough test, the sound of my voice should be at a comfortable level.

# 1

## Music, Speech, and Handedness

HOW BEING LEFT-HANDED OR RIGHT-HANDED CAN

MAKE A DIFFERENCE

⌒———————————————————————————————

The left hand has nothing to do with conducting. Its proper place is in the waistcoat pocket, from which it should emerge to restrain, or to make some minor gesture—for which in any case a scarcely perceptible glance should suffice.

—RICHARD STRAUSS[1]

THE DOOR OPENED, and the nurse wheeled Mrs. Dobkins into the lab. She was a sweet-looking lady of 64, pleasantly dressed, with her hair neatly arranged. She had agreed to be part of a study to explore how people with various types of aphasia (who had lost part or all of their ability to speak) heard musical patterns. A few weeks earlier she had suffered a stroke, produced by a clot that had lodged in the left side of her brain. This had caused some paralysis of the right side of her body, and had also seriously damaged her ability to talk. Although she could say very little, she could understand what was being said, and could communicate through gestures, using her left hand.

I introduced myself, and she nodded.

"Mrs. Dobkins," I said. "I understand that you are having some difficulty speaking."

Her eyes met mine—intense, imploring. "I would want—" she said haltingly, and then she stopped. I started to explain the study to her, but she interrupted me and grasped my hand. "I would want—I would want—I would want—"

Thoughts of the study flew away. What malicious quirk of fate had caused this sweet lady to become so terribly impaired? If only the clot had lodged in the right side of her brain instead, she would most likely have suffered very little loss of speech at all. I learned from her nurse that she had once been an English teacher, so it occurred to me that, with some luck, she might be able to recite a passage that she had learned before her stroke. (People with aphasia can sometimes sing tunes and recite passages that they had memorized before their illness.)

"Do you like Shakespeare?" I asked her.

She nodded vehemently.

"Do you know this?"

And I began to recite the beautiful lines from *The Tempest*:

> Our revels now are ended. These our actors
> As I foretold you, were all spirits, and
> Are melted into air, into thin air.

And as I continued reciting, she began, haltingly, to join in—first just a word here and there, but in the end continuously, so that finally we together recited the immortal ending:

> We are such stuff
> As dreams are made on, and our little life
> Is rounded with a sleep.[2]

Her eyes filled with tears as she realized that she could, in this way at least, produce speech. Only a short time had elapsed since her stroke, so it seemed to me that her chances of recovery were good. We proceeded with the music experiment, and as she was leaving, she again grasped my hand, and with her eyes fixed on mine, repeated "I would want—I would want—I would want—" As she was wheeled out, I prayed fervently that her wish would come true.

The average human brain weighs only three to four pounds, but it is the most complex structure in the known universe. It contains about a hundred billion nerve cells, or *neurons*. These are richly interconnected, and it is estimated that the brain involves over a hundred trillion neural connections. Given the complexity of this vast neural network, the attempt to decipher how it produces speech, perception, memory, emotion, and other mental functions presents a formidable challenge.

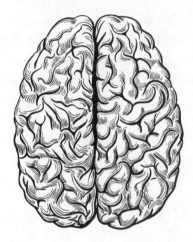

FIGURE 1.1. The left and right hemispheres of the cerebral cortex.

When you view a picture of a human brain, the first thing you notice is that it appears to be left-right symmetrical (see Figure 1.1). Most of it is composed of two walnut-shaped halves, the cerebral hemispheres, which are connected together by a massive band of fibers, known as the corpus callosum. The hemispheres are covered by a thin, heavily corrugated sheet of neurons, six layers thick, known as the cerebral cortex, and it is here that most of the activity responsible for higher mental functions occurs. Each hemisphere is divided into four lobes—the frontal, temporal, parietal, and occipital lobes, as shown in Figure 1.2.

Although our brains appear at first glance to be symmetrical, many mental functions are carried out by neural circuits that are located preferentially in one hemisphere or the other. Speech is the most dramatic example of such cerebral

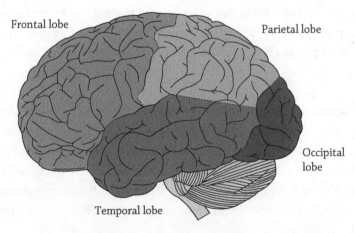

FIGURE 1.2. The lobes of the brain.

asymmetry. In most people, when a particular region of the left hemisphere is damaged, speech may become seriously impaired, whereas damage to the corresponding region of the right hemisphere might well produce only subtle derangements of speech.

It took a long time for scientists to conclude that the two hemispheres are not simply mirror images of each other. The final realization emerged out of a long and bitter controversy between those who believed that all parts of the cerebral cortex are responsible for all cortical functions, and those who argued instead that the cortex consists of a mosaic of specialized "organs," each of which occupies a particular anatomical region.

The idea of cortical specialization of function first received serious attention through the writings and lectures of the Viennese physician Franz Joseph Gall, at the turn of the nineteenth century (Figure 1.3). Gall argued that the mind is composed of many distinct faculties, each of which has a particular seat, or "organ," in the brain. He further argued that the shape of the brain is determined by the development of its various organs, and that the more powerful an organ, the larger is its size. Gall also believed that the shape of the brain determines the shape of the skull that covers it, so that one can deduce a person's aptitudes and personality characteristics by examining the pattern of bumps on his skull. Altogether,

FIGURE 1.3. Franz Joseph Gall.

he claimed that he could "read" twenty-seven mental faculties, including such abilities as language, musical talent, arithmetic, and mechanical skill, and also personality traits such as parental love, valor, and a tendency to commit larceny, and even murder.

Gall amassed what he saw to be a considerable amount of supporting evidence for his theory, by examining the skulls of people who had been particularly talented in some area, or who had extreme personality traits. In the process, he collected a large museum, in which he displayed the skulls of people who had shown various mental characteristics.

Gall's disciple, Johan Spurzheim, further developed his system (Figure 1.4), which he termed *phrenology*, and the two, together with other enthusiasts, promoted their ideas extensively through writings and public lectures, so that for a while they enjoyed enormous popularity. Political and social scientists appealed to

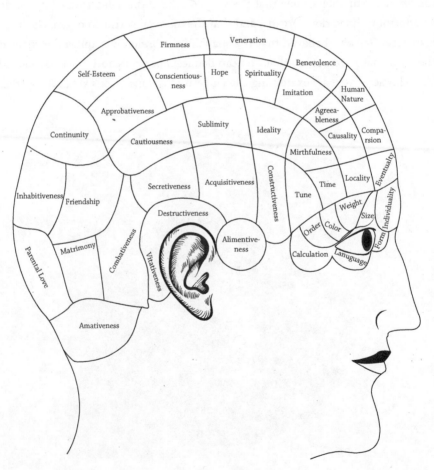

FIGURE 1.4. The organs of the brain, as depicted by phrenologists.

phrenology to support their ideas. Members of the general public took to visiting phrenologists to obtain all kinds of advice, such as whom to marry, or whether to hire a prospective employee. For a while, phrenologists held a position in the public mind that was very similar to the position horoscope readers hold today.

But scientifically, the phrenologists overreached themselves through their extravagant and poorly founded claims, and the inevitable backlash occurred. The French physiologist Pierre Flourens conducted a number of experiments on animals, from which he concluded that when various regions of the cortex were destroyed, the different mental functions were affected together, and not independently. He also launched a vicious personal attack on Gall, portraying him as a lunatic who was obsessed with collecting peoples' skulls. As he wrote,

> At one time everybody in Vienna was trembling for his head, and fearing that after his death it would be put in requisition to enrich Dr. Gall's cabinet. . . . Too many people were led to suppose themselves the objects of the doctor's regards, and imagined their heads to be especially longed for by him.[3]

Opinion against the phrenologists grew, and finally their scientific respectability disappeared. But the concept of cerebral localization of function did not die, and thoughtful scientists turned to the study of patients with brain damage in attempts to verify it. Prominent among these scientists were Jean-Baptiste Bouillard, who argued from his observations that speech and language were localized in the frontal lobes. Bouillard did not differentiate between the left and right frontal lobes, assuming (as did the phrenologists) that the brain regions responsible for the different mental functions were located symmetrically in the two hemispheres. As a result, convincing counterexamples to his claims were reported, and the controversy continued to rage.

Enter Pierre Paul Broca (Figure 1.5), the brilliant French surgeon and physical anthropologist who had recently founded the esteemed Société d'Anthropologie in Paris. A careful and thorough researcher, Broca became increasingly drawn to the localizationist view. It so happened that a patient named Leborgne had recently been transferred to his surgical service at the Bicêtre Hospital in Paris. This patient had for many years lost his power of speech, and could only utter a single syllable, "tan," with the result that he was known as "Tan" by all in the hospital. Ultimately he became paralyzed, and on April 17, 1861, he died. The next day, Broca performed an autopsy, and immediately delivered a report to the society that he had founded. The autopsy had shown a massive softening of the left frontal lobe, particularly in an area toward the back of this lobe—now known as Broca's

FIGURE 1.5.  Dr. Pierre Paul Broca.

area—and Broca claimed that this lesion had been responsible for Tan's loss of speech. A few months later, Broca reported the results of an autopsy on another patient who had lost his power of speech, which also revealed a lesion in the left frontal lobe.

During the next couple of years, more cases were drawn to Broca's attention, and in a further lecture he described eight cases of loss of speech, all associated with damage to the frontal lobes of the left hemisphere. There were also reports of patients who had corresponding lesions in the frontal lobe of the right hemisphere, but whose speech was not impaired. Taking all this evidence together, in 1865 Broca finally made his famous pronouncement: "We speak with the left hemisphere," and the theory of speech localization of function, coupled with the concept of hemispheric asymmetry, became widely accepted.

Soon after, in 1874, the German scientist Carl Wernicke identified another region of particular importance to speech—this is also in the left hemisphere, but in the temporal lobe. Whereas people who had lost their speech due to damage to Broca's area (such as Mrs. Dobkins) can often understand speech quite well, their spontaneous speech is typically slow, and consists of only a few words. In contrast, people with damage to Wernicke's area have difficulty understanding what is being said, and their speech, while fluent, sounds rambling and nonsensical.

꧁

Viewed in the flesh, the left and right sides of our bodies appear as mirror images of each other. The bilateral symmetry of our external bodies is associated for the most part with a corresponding symmetry of the brain structures that control our movements. For reasons that are mysterious, the two sides are switched around, so that the left hemisphere controls the right side of the body, and the right hemisphere controls the left side. (Similarly, the left hemisphere registers locations that are seen and heard in the right side of space, and the right hemisphere registers locations that are seen and heard in the left side.) So when we use our right hand for writing, our left hemisphere is directing its movements.

Although the two sides of our bodies appear symmetrical when viewed at rest, we do not move our limbs symmetrically. Throughout the ages, and in all cultures that have been studied, most people have favored their right hand. Experts have examined works of art depicting people carrying tools and weapons, and have concluded that the large majority of humans have been right-handed over a period of at least fifty centuries. Other experts have deduced, based on examination of prehistoric tools, that the tendency for right-handedness stretches as far back as *Australopithecus*, who lived around three and four million years ago.

Earlier I had said that the human brain appears at first glance to be bilaterally symmetrical. But on closer inspection, the left and right hemispheres do have differences that are visible from the outside, and these differences correlate with handedness. In right-handers the planum temporale, a region in the temporal lobe that is part of Wernicke's speech area, is generally larger on the left than on the right. This asymmetry occurs early in development, and has even been found in unborn babies. It also goes back far into human prehistory. The shape of the intracranial cavity is determined by the shape of the brain that it covers, so it is possible to infer the shape of the brain in prehistoric humans by examining casts made of the insides of skulls. Using this method, researchers have found that the skulls of Neanderthals, who lived about 50,000 to 60,000 years ago, show asymmetries that are similar to those in modern right-handers. This is even true of the skulls of Peking Man, *Homo erectus pekinensis*, who lived from roughly 300,000 to 750,000 years ago, and similar asymmetries have been found in the great apes. But as would be expected, this asymmetry is far less striking among left-handers, and they sometimes have a larger planum temporale on the right side of their brain instead of their left side.

Right-handers tend to be right-eared, though we only notice this under special circumstances. A strongly right-eared and right-handed person will go to great lengths and even inconvenience to use both their right ear and right hand at the same time. When the telephone rings, I pick it up with my right hand, and I transfer it to my left hand. Then with my left hand I put the phone to my right ear, which leaves my right hand free for writing. If I need to put the phone to my left ear instead, speech coming from the telephone receiver sounds strange, even though on hearing tests my left ear performs better than my right one.

Because the left hemisphere is important to both speech and also movement of the right side of the body, loss of speech resulting from brain injury is often accompanied by some paralysis of the right side. Knowledge of the association between aphasia and right sided paralysis extends far back into the past, and is reflected in the powerful words from Psalm 137:

> If I forget thee, O Jerusalem, let my right hand forget its cunning
>
> If I do not remember thee, let my tongue cleave to the roof of my mouth

Now although most people are strongly right-handed, handedness really forms a continuum—from strong right-handers to mixed right-handers to mixed left-handers to strong left-handers. There are a number of tests to determine where a person stands on this continuum. I use two, one called the Edinburgh Handedness Inventory,[4] and the other developed by Varney and Benton.[5] The second one is shown here, and you might want to test your handedness by filling it out. People are designated as strong right-handers if they score at least nine "Rights" on this questionnaire, and as strong left-handers if they score at least nine "Lefts"; they are designated as mixed-handers if they score anything in between.

၇

| | | | |
|---|---|---|---|
| 1. With which hand do you write? | Right | Left | Either |
| 2. With which hand do you use a tennis racquet? | Right | Left | Either |
| 3. With which hand do you use a screwdriver? | Right | Left | Either |
| 4. With which hand do you throw a ball? | Right | Left | Either |
| 5. With which hand do you use a needle in sewing? | Right | Left | Either |
| 6. With which hand do you use a hammer? | Right | Left | Either |
| 7. With which hand do you light a match? | Right | Left | Either |
| 8. With which hand do you use a toothbrush? | Right | Left | Either |
| 9. With which hand do you deal cards? | Right | Left | Either |
| 10. With which hand do you hold a knife when carving meat? | Right | Left | Either |

People's scores sometimes surprise them. Once I was chatting with a bright engineering student who was evidently left-handed. During our conversation she spotted the questionnaire and asked if she could fill it out. To my surprise she circled all the answers as "Right," including "With which hand do you write?" but she wrote her answers down with her left hand!

"Karen," I asked, "why did you write with your left hand that you write with your right hand?"

"Well," she answered, that's because I do."

"But you just used your left hand to fill out the questionnaire."

"I know," she replied "but that's because I felt like it."

"So how often do you feel like using your left hand, then?"

"Almost all the time. But I'm right-handed really."

It's amazing to discover how many variations there are in the ways people fill out these questionnaires. You can get any pattern of responses. Yet if you query people about them, they often argue that the answer they gave to each question was the only logical one. I circle "Right" to all the questions except for "With which hand do you deal cards?" Here I circle a definite "Left," and find it hard to understand how anyone would want to deal cards with their right hand. Yet most strong right-handers circle "Right" in answer to this question.

Most interesting are those rare people who are truly ambidextrous, and circle "Either" in answer to all or most of the questions. They use their right hand to comb their hair, shave, and brush their teeth on their right sides; and they use their left hand to do these things on their left. They throw a ball or swing a tennis racket with either hand, depending on the circumstances. If logic has anything to do with it, perhaps ambidextrous people are the most logical of all.

The relationship between handedness and speech is complex. In most right-handers the left side of the brain is particularly important for speech, and we call them "left hemisphere dominant." Two-thirds of left-handers follow this pattern, but in one-third the right hemisphere is more important for speech instead. Also in many left-handers, both hemispheres play a significant role. This is reflected in statistics on recovery from stroke. The chances of recovery are considerably better for a left-hander, since when one side of the brain is damaged, the speech areas in the undamaged side may take over.

There is yet another complication. If someone has a left-hander in his immediate family (a parent or sibling), it is more likely that both hemispheres play a significant role in his speech. So as would be expected, right-handers are more likely to recover their speech following a stroke if their family includes a left-hander. The Russian neurologist Alexander Luria studied patients who had suffered penetrating wounds in their speech areas that resulted in aphasia. After some time had elapsed following the injury, many of the patients made a good recovery, and this group contained a large proportion of left-handers as well as right-handers with left-handed relatives.[6]

ᒍ

Left-handers get a mixed press from psychologists who discuss mental abilities. A number of researchers have argued that there is a higher incidence of dyslexia and learning disabilities among left-handers. Others have found considerably larger numbers of left-handers and mixed-handers among students who major in art compared with those who major in other fields. In one study, left- or mixed-handers comprised 47 percent of art students at the Massachusetts College of Art and the Boston University School of Fine Arts, compared with 22 percent of undergraduate students of liberal arts at Boston University.[7] An impressive number of great artists were left-handed, including Michelangelo, Leonardo da Vinci, Raphael, Hans Holbein, Paul Klee, and Pablo Picasso. M. C. Escher wrote with his right hand and drew with his left hand. He wrote,

> meanwhile I am annoyed that my writing should be so shaky; this is due to tiredness, even in my right hand in spite of the fact that I draw and engrave with my left. However, it seems that my right hand shares so much in the tension that it gets tired in sympathy.[8]

Perhaps this served as inspiration for his lithograph, "Drawing Hands," shown in Figure 1.6.

The issue of music is particularly complicated, since most musical instruments are designed for right-handers, and so can be difficult for left-handers to play. Keyboard instruments such as the piano and organ are so arranged that the right hand generally plays the more prominent part as well as the more challenging one. Stringed instruments, such as the violin and the guitar, also pose a difficulty for left-handers. Here the right hand does the bowing or plucking, and so is responsible for producing the sound, whereas the left hand "sets the stage" by placing the fingers in different positions along the string. Dedicated musicians who are strongly left-handed sometimes "remake" their instruments so that they can play them in reverse. Charlie Chaplin shifted the bar and soundpost, and restrung his violin, so that he could play it left-right reversed. (You can see him playing the violin this way in the movie *Limelight*.) In the pop music world, the bass player Paul McCartney and guitarist Jimi Hendrix , who were both left-handed, used right-handed instruments, but turned them over and restrung them.[9]

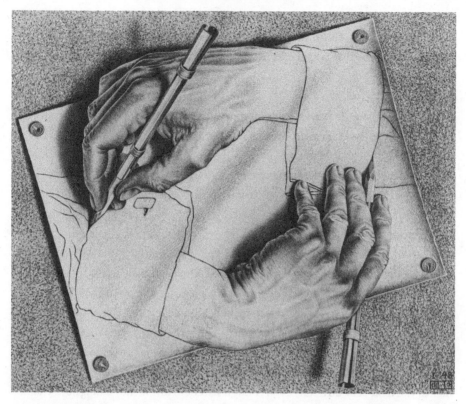

FIGURE 1.6. M. C. Escher's *Drawing Hands*. © 2018 The M. C. Escher Company—The Netherlands. All rights reserved. www.mcescher.com. It is a "Strange Loop" (see Chapter 4).

Left-handed string players also have problems when they perform in an orchestra: If they bow with their left hand, they risk bumping into their right-handed neighbor to their left. For this reason, their presence in orchestras is quite rare—and this rarity can have unfortunate consequences. There's a story concerning the renowned Hungarian conductor János Ferencsik. Apparently he had been drinking in a tavern well into the night, and returned home just before dawn. Since he had a rehearsal appointment that morning, he took a short nap, downed a few cups of coffee, and made his way to the concert hall. Upon ascending the podium, he spotted two left-handed violinists sitting side by side! Thoroughly alarmed by what he took to be a hallucination, he announced "No rehearsal today!" and hastily departed.[10]

At all events, given the problems for left-handers posed by many musical instruments, we would expect that they would be rare among performing musicians. We might expect conductors to be an exception, since either hand can wield the baton. But left-handed conductors encounter a bias against left-handedness also. Richard Strauss, who is best known as a composer but was also a conductor, put it like this:

The left hand has nothing to do with conducting. Its proper place is in the waistcoat pocket, from which it should emerge to restrain, or to make some minor gesture—for which in any case a scarcely perceptible glance should suffice.[11]

But despite the obstacles encountered by left-handers, they don't appear to be underrepresented among musicians. Carolus Oldfield at the University of Edinburgh found that left-handers were just as prevalent among music students as among psychology undergraduates.[12]

So given the difficulties faced by left-handed instrumentalists, we might surmise that left-handers would win out statistically over right-handers where perception and memory for music are concerned. My own involvement with this question arose from a chance observation. I was carrying out experiments on memory for musical pitch, using the following procedure: First you play a test tone, and this is followed by six additional tones, then by a pause, and then by a second test tone. The test tones are either the same in pitch or they differ by a semitone. The subjects listen to the first test tone, ignore the remaining tones that come before the pause, and then judge whether the second test tone is the same in pitch as the first, or different. Interestingly, having an intervening sequence of tones generally makes the task very difficult, even though listeners know they can ignore them.[13]

For one experiment, I was searching for subjects who performed very well on this task. So I would give a few examples to people who were interested in participating, and select those who made very few errors. When I had finally selected an excellent group of subjects, I found that a large number of them were left-handed. Did that mean that left-handers in general might do better on my musical memory task, I wondered?

To find out, I recruited two new groups of right-handers and left-handers, who had received roughly the same amount of musical training.[14] As I had surmised, the left-handers made significantly fewer errors than the right-handers on this task. When I further divided the groups into four—strong right-handers, mixed right-handers, mixed left-handers, and strong left-handers—I found that the mixed left-handers outperformed all the other groups.[15]

How could these results be explained? Let's suppose that in a large proportion of mixed left-handers, pitch information is stored in both hemispheres, rather than in a single hemisphere—just as in a large proportion of left-handers, both hemispheres are involved in speech. If the pitch of the first test tone were stored in two places rather than one, then the listener would have a greater chance of retrieving it, and so of producing the correct answer.

So, in a further experiment I recruited a new group of subjects, this time selecting only those who had received little or no musical training. I used a different pitch memory task. Here I played a sequence of five tones, which was followed by a pause, and then by a test tone. The subjects' task was to decide whether the test tone had been included in the sequence that occurred before the pause. When I studied the pattern of errors, the same relationships to handedness emerged. The left-handers as a group significantly outperformed the right-handers, and when the groups were further divided into four, the mixed left-handers again significantly outperformed the other three groups.[16]

So from these two experiments, it appears that people who are left-handed, but not strongly so, are likely to perform particularly well on tasks involving memory for musical pitch. This opens up a host of questions. Are left-handers as a group also better at tasks involving rhythm, or timbre, or other features of music? At present, these issues remain unexplored.

We have described the history of thought about specialization of function in the brain with respect to speech and music. Against this background we next explore a number of musical illusions that are heard differently, on a statistical basis, by right-handers and left-handers.

# 2

## Some Musical Illusions Are Discovered

~~~

IT WAS THE fall of 1973, and I was experimenting with some new software that would enable me to play two sequences of tones at the same time, one to my right ear and the other to my left one. I was expecting in this way to explore various longstanding questions about how we perceive and remember sounds. But as the afternoon wore on, it became increasingly clear that something rather strange was happening—my perceptions were not at all as I had expected. So I decided to step back and try a very simple pattern.

The pattern I devised consisted of two tones that were spaced an octave apart, and alternated repeatedly. The identical sequence was played to both ears simultaneously; however, the tones were out of step with each other, so that when the right ear received the high tone, the left ear received the low tone, and vice versa. In other words, the right ear received the sequence "high tone—low tone—high tone—low tone" over and over again, while at the same time the left ear received "low tone—high tone—low tone—high tone" over and over again. I reasoned that with this simple pattern at least, I could make sense of what I heard, and then return to more complicated examples later.

But when I put on my earphones, I was astonished by what I heard: A single tone appeared to be switching back and forth from ear to ear, and at the same time its pitch appeared to be switching back and forth from high to low. In other words, my right ear heard "high tone—silence—high tone–silence" while at the same time my left ear heard "silence—low tone—silence—low tone" (Figure 2.1).

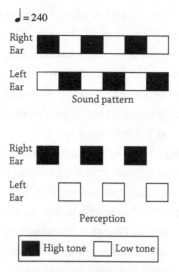

FIGURE 2.1. The pattern that produces the octave illusion, and a way it is often perceived when played through stereo headphones. Although both ears receive a high tone alternating with a low tone, most listeners hear an intermittent high tone in one ear that alternates with an intermittent low tone in the other ear. From Deutsch (1974b).

    I carefully studied the specifications I had entered into the computer—there was no mistake there. My next thought was that the software must somehow be at fault. Instinctively I reversed the positions of the earphones, expecting that the positions of the high and low tones would reverse with them. But to my amazement, this made no difference at all to my perception—my right ear continued to hear "high tone—silence—high tone—silence" and my left ear continued to hear "silence—low tone—silence—low tone." So it seemed as though reversing the earphone positions had caused the high tone to migrate from one earphone to the other, and the low tone to migrate in the opposite direction! With growing incredulity, I reversed the earphone positions over and over again, only to find that the high tone obstinately remained in my right ear and the low tone in my left one.

    The next thing that occurred to me was that something really peculiar had happened to my hearing. Somewhat alarmed, I grabbed a few people from the corridor to find out whether they, at least, heard the pattern as would have been expected. To my amazement (as well as my relief), most of them experienced the same illusion as I did, and none could guess what was really being played.

It is difficult to convey my thoughts when I was finally convinced that I had discovered a real, and utterly paradoxical, illusion.[1] There was nothing in the scientific literature or in my own experience that could begin to explain this bizarre effect. I was astounded to realize that our hearing mechanism, in other ways so cleverly and intricately constructed, could go so wildly wrong when faced with such a simple pattern. I also realized that this must surely be one of a new class of illusions, whose basis was as yet unknown.

What was going on here? I knew that under the right conditions, we can hear sounds that are so low in amplitude that they involve an amount of air displacement corresponding to less than the diameter of a hydrogen atom. We can detect time differences between the onsets of sounds at the two ears that are less than ten millionths of a second. We can remember long pieces of music with extraordinary accuracy, and can detect sophisticated relationships between musical passages with ease. How could the auditory system be capable of these amazing feats of analysis, and yet come up with such an absurd interpretation of this trivial pattern?

The *octave illusion* (as I named it) presents us with a logical conundrum. We can explain the perception of alternating pitches by assuming that the listener hears the sounds arriving at one ear and ignores the others. But then, both the high and the low tones should appear to be arriving at the same ear. Or, we can explain the alternation of a single tone from ear to ear by supposing that the listener switches his attention back and forth between ears. But then the pitch of the tone that he hears should not change with the change in the tone's apparent location—the listener should hear either a high tone that alternates from ear to ear, or a low tone that alternates from ear to ear. The illusion of a single tone that alternates *both* in pitch *and* in location is quite paradoxical.

The illusion is even more surprising when we consider what happens when the earphones are placed in reverse position. Now most people hear *exactly the same thing*: The tone that had appeared in the right ear still appears in the right ear, and the tone that had appeared in the left ear still appears in the left ear. This produces the unsettling impression that reversing the earphone positions causes the high tone to migrate from one earphone to another, and the low tone to migrate in the opposite direction! You can experience this in the "Octave Illusion" module.

The illusion has yet another surprising aspect: Right-handers and left-handers differ statistically in terms of where the high and low tones appear to be coming from. Right-handers tend strongly to hear the high tone on the right and the low tone on the left, but left-handers as a group are quite divided in terms of what they hear. In one experiment, I separated my listeners into three groups—strong right-handers, mixed-handers, and strong left-handers. Then I further separated each group into two: those who had a left- or mixed-handed parent or sibling, and those who had only right-handed parents and siblings.

This six-way division produced striking results: Strong right-handers were more likely to hear the high tone on the right than were mixed-handers, and mixed-handers were more likely to do so than were left-handers. And for each handedness group, those with only right-handers in their family were more likely to hear the high tone on the right than were those with left- or mixed-handers in their family.[2] So given the relationships between patterns of cerebral dominance and handedness that I discussed earlier, it appears that in listening to the octave illusion, we tend to hear the high tone in the ear that is opposite our dominant hemisphere—in right- handers, this is usually the right ear—and that we tend to hear the low tone in the opposite ear.

But how, more specifically, can we account for this illusion? Let's assume that there are two separate systems in the brain, one determining *what* pitch we hear and the other determining *where* each tone is coming from. This puts us in a position to advance an explanation. To determine *what* we hear, we attend to the pitches arriving at our dominant ear, and suppress from consciousness those arriving at our nondominant ear. (So most right-handers attend to the pitches arriving at their right ear rather than their left one.) On the other hand, the *where* system follows an entirely different rule: Each tone appears to be arriving at the ear that receives the higher tone, regardless of whether the higher or the lower tone is in fact perceived.

To see how the model works, let's first consider a right-eared listener. When a high tone is delivered to his right ear and a low tone to his left, he hears a high tone, since this is delivered to his right ear. He also localizes the tone in his right ear, since this ear is receiving the higher tone. But when a low tone is delivered to his right ear and a high tone to his left, he now hears a low tone, since this is delivered to his right ear—but he hears the tone in his left ear instead, since this ear is receiving the higher tone. So he hears the entire sequence as a high tone to his right that alternates with a low tone to his left. You can see that reversing the earphone positions would not alter this basic percept—the sequence would simply appear to be offset by one tone. But when we consider the model for a left-eared listener instead, we see that he should hear the identical pattern as a high tone to his left alternating with a low tone to his right (Figure 2.2). And a number of experiments have confirmed this model.[3]

The octave illusion therefore provides a clear example of what psychologists call an *illusory conjunction*. In listening to this repeating pattern, a typical right-eared listener, when presented with a low tone to the right and a high tone to the left, combines the pitch of the tone to the right with the location of the tone to the left, and so *perceptually creates a tone that does not in fact exist!*

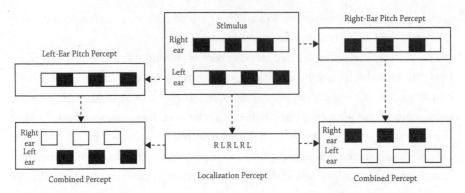

FIGURE 2.2. Model showing how the outputs of two decision mechanisms, one determining perceived pitch and the other determining perceived location, can combine to produce the octave illusion. The black boxes indicate high tones and the white boxes indicate low tones. From Deutsch (1981).

Illusory conjunctions also occur with visual patterns. Ann Treisman and her colleagues have found that when viewers were shown several colored letters simultaneously, they sometimes combined the colors and shapes of the letters incorrectly. For example, when they were presented with a blue cross and a red circle, they sometimes reported seeing a red cross and a blue circle instead.[4] But to produce such illusory conjunctions in vision, the viewer's attention needs to be overloaded in some way; for example, by having him perform a different task at the same time. Yet in the octave illusion—and in other illusory conjunctions involving music that I'll be describing later—the presented sounds are very simple, and the viewer's attention is not overloaded. So attention limitations cannot explain the illusory conjunctions that underlie the octave illusion.

Sometimes, as people continue listening to the octave illusion, the high and low tones suddenly appear to reverse position. The high tone may first be heard on the right and the low tone on the left. Then after a few seconds, the high tone suddenly switches to the left and the low tone to the right. Then after a few more seconds, the tones interchange positions again. And so on. These sudden perceptual reversals are like reversals of ambiguous figures in vision. Take the Necker cube shown in Figure 2.3. One way you might see this cube is with its front to the lower left and its back to the upper right. Alternatively, you might see it with its front to the upper right and its back to the lower left. If you fix your eyes on the cube for a while, you will find that your perception suddenly flips without warning, so that the back face suddenly becomes the front one. Then it flips back again, then it returns to its original position, and so on, back and forth. But you can't view the cube in both these orientations simultaneously.

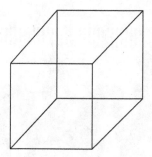

FIGURE 2.3. The Necker cube. In viewing this figure, the back face of the cube periodically becomes the front one. This visual illusion is analogous to some percepts of the octave illusion, in which the high and low tones appear periodically to exchange locations.

The nineteenth-century crystallographer Louis Albert Necker, a Professor of Minerology at Geneva, after whom the Necker cube is named, remarked in 1832:

> The object I have now to call your attention to, is an observation . . . which has often occurred to me while examining figures and engraved plates of crystalline forms: I mean a sudden and involuntary change in apparent position of a crystal or solid represented in an engraved figure. (p. 336)[5]

The Dutch artist M. C. Escher has created many drawings that produce similar perceptual reversals. Take a look at his *Regular Division of the Plane III*, shown in Figure 2.4.[6] In the upper part of this picture, the black horsemen clearly provide the figure and the white horsemen the ground. In the lower part, the white horsemen instead provide the figure, and the black horsemen the ground. But in the middle there is a region of ambiguity in which you can perceive the horsemen either way. When you fix your eyes on this middle region, you find that your perception alternates between the two possibilities.

What happens when the sounds are presented through loudspeakers rather than earphones? In one experiment, which I carried out in an anechoic chamber (a room that is specially treated so as to be free of echoes), the listener was positioned so that one loudspeaker was exactly on his right and the other loudspeaker exactly on his left. When I played the octave illusion, the listener heard a high tone as though coming from the speaker on his right, and this appeared to alternate with a low tone that appeared to be coming from the speaker on his left. When the listener slowly turned around, the high tone continued to appear on his right and the low tone on his left. This perception continued until the listener was facing one speaker with the other speaker directly behind him. The illusion then abruptly disappeared, and instead the listener heard a blend of the high and low tones as

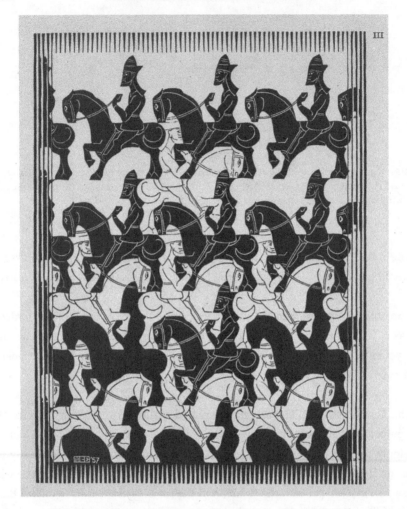

FIGURE 2.4. M. C. Escher, *Regular Division of the Plane III*. © 2018 The M. C. Escher Company—The Netherlands. All rights reserved. www.mcescher.com.
In the upper part of the woodcut, the black horsemen provide the figure and the white horsemen the ground. In the lower part, the white horsemen provide the figure and the black horsemen the ground. In the middle part, your perception flips back and forth between these two alternatives.

though they were simultaneously coming from both speakers. But as he continued turning, the illusion abruptly reappeared, with the high tone still on his right and the low tone on his left. So when he had turned by 180 degrees, it appeared to him that the speaker that had been producing the high tone was now producing the low tone, and the speaker that had been producing the low tone was now producing the high tone!

To give you a sense of how unsettling this experience can be, here's how an analogous illusion in vision would appear. Suppose you are standing in a room, looking straight ahead. Your eyes take in a chair to your right, a table in front of you, a window to your left, and so on. Then as you rotate slowly, the chair appears to rotate along with you, as does the table, the window—in fact, the entire room appears to be rotating with you. So when you have turned around completely, the chair, table, and window all appear to have moved to the opposite end of the room, and to be facing in the opposite direction!

The well-known ventriloquism illusion is also based on the fact that our auditory *what* and *where* mechanisms follow different rules. When a ventriloquist induces an audience of delighted children to believe that he is carrying out a conversation with a dummy, he achieves this effect by making the dummy's mouth move in synchrony with his speech, while keeping his own lips as motionless as possible. So the children in the audience correctly identify what speech is being produced, but are fooled by their visual impressions into believing that the speech is coming from the dummy. We experience similar illusions when we are watching a movie, and see an actor's lips move in synchrony with his speech sounds, or see horses galloping across the screen in synchrony with the sound of pounding hoofs.

Another illusion that illustrates the separation of *what* and *where* mechanisms is known as the *precedence effect*. To experience this illusion, sit facing two loudspeakers that are some distance from each other, with one to your right and the other to your left. The identical stream of speech or music is played through the two speakers; however, the sounds coming from the speakers are slightly offset in time. When the duration between the onsets of the two streams is less than about 30 ms, you hear only a single stream, which seems to be coming from the speaker that produces the earlier-arriving sound. The other speaker appears to be silent, even if the sound it is producing is louder.[7]

The precedence effect is useful since it suppresses unwanted echoes and reverberation from conscious perception. It's employed in the design of sound systems in auditoria and concert halls. So, for example, when someone is lecturing from a podium and you are in the audience some distance away, their voice might be difficult for you to follow. But if loudspeakers are placed in the walls around the audience, you can hear the lecturer's voice well—the sound you are hearing is coming from a loudspeaker near you, but it appears to be coming from the podium, and not from the loudspeaker. The composer Michael A. Levine has pointed out to me that, in the case of music, delaying the output from the two loudspeakers makes a mono sound appear "bigger"—and the longer the delay (within 30 ms or so, with the amount of delay determined by the tempo of the music), the greater is this effect.

A further intriguing *what–where* dissociation was created by Nico Franssen in the Netherlands. He set up two loudspeakers in front of the listener—with one to his left, and the other to his right. He then presented a very brief sound (a click) through the loudspeaker to the left, while he gradually turned on a long, steady tone, which he presented through the loudspeaker to the right. The listener heard both sounds, but, amazingly, he heard them as a single sound that began with the click and continued with the long tone, and which appeared to be coming from the left loudspeaker only. Yet the left loudspeaker was in reality silent while the long tone was sounding, and the right loudspeaker, which was actually producing the tone, appeared to be silent instead. So the listener heard the long tone correctly, but heard it in the wrong location![8] The physicist Manfred Schroeder has told me that when he presents this demonstration in a strongly reverberant room, this illusion can be so strong that he sometimes has to unplug the cord to the left loudspeaker and hold it up for listeners to see, in order to convince them that this loudspeaker is not really producing any sound.

Recently, Harvard physicist Eric Heller sent me an email describing another remarkable *what–where* dissociation. Using earphones, he first presented a complex tone to both ears, and delayed the onset of the tone to the left ear by 1 ms. As would be expected, this delay caused the tone to be heard as in the right ear, with no sound being heard as in the left ear. Heller then applied a tremolo (amplitude modulation) to the tone in the left ear, and surprisingly, this caused the tremolo to sound loud and clear, but in the *right* ear, while the left ear continued to hear only silence! So in this example, which Heller named the *tremolo transfer illusion, where* the sound was heard continued to be on the right, but *what* was heard resulted from the input to the left![9]

Bizarre *what–where* dissociations can occur in patients who are suffering from "unilateral neglect"—a condition that often follows a stroke in the right hemisphere, particularly when the parietal lobe is affected. Remember that the right hemisphere registers sights and sounds in the left half of space. Neglect patients don't notice objects or events that occur in the left part of their world. When food is placed in front of them, they notice only the portion on their right, and they deny seeing the portion on their left—even complaining that they haven't been given enough to eat! They comb their hair and brush their teeth on their right side only, leaving their left side untended. They bump into objects on their left as they move around, since they don't notice them. And—most importantly for our purposes— they only notice sounds that come from their right, and they deny hearing sounds that come from their left. It isn't that they are blind or deaf for things on their left—they just don't notice them, and can't even make themselves notice them.

Leon Deouell and Nachum Soroker at Tel Aviv University in Israel devised an ingenious experiment to show that patients with unilateral neglect could be induced to hear sounds that were coming from their left, by duping them into believing that they were instead coming from their right.[10] They presented a dummy loudspeaker to the right of the patient, where he could see it, and then presented sounds through a hidden loudspeaker that was positioned to his left. Fooled by the presence of the dummy loudspeaker, the patients with unilateral neglect reported hearing more of the sounds that came from their left—but they insisted that these sounds were coming from their right instead! In this way, they were induced to report correctly *what* sound was presented, though they reported incorrectly *where* the sound was coming from.

So what brain circuitry is responsible for these extraordinary *what–where* dissociations? Joseph Rauschecker, Biao Tian, and their colleagues at Georgetown University have studied the ways in which neurons in the monkey's brain respond to different aspects of sound. In particular, they examined the responses of neurons in an area called the lateral belt, which lies next to the primary auditory cortex. They found that neurons in one part of the lateral belt—the anterior belt—responded strongly to monkey calls, and different neurons in this area responded specifically to different types of call. Neurons respond to signals by increasing or decreasing their firing rate. So when a particular monkey call sounded, a lateral belt neuron might fire off like a machine gun, but the same neuron would fire rather weakly in response to different monkey calls. In general, lateral belt neurons were not fussy about *where* the sounds were coming from, provided that the right type of monkey call was presented. In contrast, neurons in another part of the lateral belt—the caudal belt—responded strongly to calls that came from particular spatial locations, but not others—and many of these neurons were not fussy about *what* type of call was presented.[11]

So in a nutshell, neurons in the anterior belt responded selectively to *what* sound was being presented, whereas those in the caudal belt responded selectively to *where* the sound was coming from. In further studies, the researchers showed that neurons from these two areas send signals to different parts of the brain, in such a way as to form two separate streams—a *what* stream and a *where* stream. And importantly, in patients with unilateral neglect, the *where* stream involved the parietal lobe. So it makes sense that patients with damage to this brain region should have problems dealing with spatial aspects of sound, while at the same time being able to identify the sounds correctly.

❧

My discovery of the octave illusion led me to wonder what other illusions might exist that have a similar basis. The night following this discovery, I kept hearing, in my mind's ear, a multitude of tones that danced around me, weaving fantastic patterns in space. Gradually they crystallized into the simple repeating pattern shown in the upper part of Figure 2.5. Early the next

**Scale Illusion**

http://dianadeutsch.net/ch02ex02

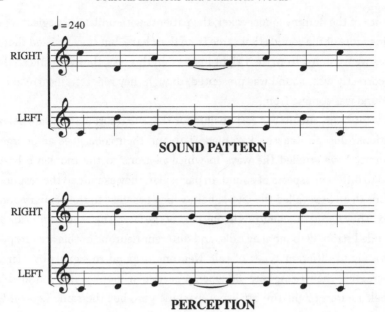

FIGURE 2.5. The pattern that produces the scale illusion, and a way it is often perceived when listening through headphones. From Deutsch (2013a).

morning, I scribbled the sequence of notes down on a piece of scrap paper, headed back to the lab, and entered it into a computer. Amazingly, another illusion appeared just as I had imagined it—I heard all the higher tones in my right ear and all the lower tones in my left ear, as shown in the lower part of Figure 2.5. When I switched the earphones around, the higher tones remained on my right and the lower tones on my left. This again produced the bizarre impression that the earphone that had been emitting the higher tones was now emitting the lower tones, and vice versa. And instead of hearing the two jagged sequences that were being played, I heard two smooth melodies, a higher one and a lower one, that moved in opposite directions. My mind was rejecting the jerky and improbable sequences that were being spewed out of the computer, and reorganizing the tones so as to make musical sense. The pattern I heard, which I named the *scale illusion,* is shown in Figure 2.5, and you can hear it in the "Scale Illusion" module.[12]

At this point you may be wondering whether this could be a mere laboratory curiosity, in which the brain is made to arrive at the wrong conclusions under highly unusual circumstances. Yet during a lecture that I gave at the University of California, Irvine, in dedication of a newly renovated concert hall, we played the scale illusion with three violinists on the extreme left of the stage, and three on the extreme right. People in the audience—even those who were sitting to one side—experienced the illusion very clearly, so that half the notes appeared to be coming from the wrong spatial locations.

Related illusions can also be created with other musical passages that are designed along the same lines, so that, for example, the brain may assign a theme to one location in space, and its accompaniment to the other, regardless of how the two parts are being played. In fact, an illusion similar to the scale illusion may have been the cause of a serious disagreement between the composer Pyotr Ilyich Tchaikovsky and the conductor Artur Nikisch at the end of the nineteenth century. In the summer of 1893, Nikisch met with Tchaikovsky to discuss his Sixth Symphony (the *Pathétique*), which Nikisch later conducted. The last movement of the symphony begins with a passage in which the theme and accompaniment waft back and forth between the two violin parts. In other words, the first violins play one jagged sequence while the second violins play a different, but overlapping, jagged sequence. I should add that at the time, instruments were so arranged that the first violins were to the left of the orchestra and the second violins to the right, so that they were clearly separated in space. (Today the two sections are generally placed next to each other, so you would expect their sounds to blend.)

For some reason, Nikisch took strong exception to Tchaikovsky's scoring, and tried to insist that he rescore the passage so that the first violins would play the theme and the second violins the accompaniment. However, Tchaikovsky positively refused to agree to this change, and the piece was premiered as originally written. Yet Nikisch continued to feel so strongly that he rescored the passage anyway, and so initiated an alternative tradition of performing the symphony.

**Tchaikovsky Passage**

http://dianadeutsch.net/ch02ex03

A few conductors still follow Nikisch's rescored version, though most adhere to Tchaikovsky's original one (see Figure 2.6).

Nobody knows why these two great musicians disagreed with each other so strongly here, and there is no evidence that either of them realized that they were dealing with an illusion. Some people have speculated that Nikisch wanted to extract the best performance out of the orchestra—his rescored version is easier to play—while Tchaikovsky was determined that the audience should hear the notes from the theme waft back and forth across the stage. Another interpretation, suggested by the composer Michael A. Levine, is that the spatial separation of the violin parts makes the phrase sound richer and "bigger"— a holistic effect that Tchaikovsky may have been trying to produce. However,

FIGURE 2.6. The beginning of the final movement of Tchaikovsky's Sixth Symphony—*The Pathétique*. The upper part of the figure shows the passage as it is played by the two violin sections, and the lower part shows how it is generally perceived. From Deutsch (2013a).

there is no doubt that this passage produces an illusion—the theme is heard as though coming from one set of instruments and the accompaniment as from the other. I experienced this effect very strongly when the John Angier Production Company came to film my lab for an episode of *NOVA*, and they filmed the UCSD Symphony performing this passage with the orchestra arranged as in Tchaikovsky's time.[13] The recording they made is in the "Tchaikovsky Passage" module.

Other examples of this type of spatial reorganization occur in a broad range of musical situations. The music theorist David Butler played melodic patterns to music students through spatially separated loudspeakers, and they notated what they heard as coming from each speaker. Virtually all the notations consisted of higher and lower melodic lines, rather than the patterns that were presented. Differences in timbre were sometimes introduced between the sounds coming from the two speakers, and this caused the subjects to hear a new tone quality, but it appeared to be coming simultaneously from both the left and the right speakers.[14,15]

Another example, which is described in John Sloboda's influential book *The Musical Mind*,[16] occurs at the end of Rachmaninoff's *Second Suite for Two Pianos*. Here the first and second pianos play patterns involving the same two tones. Yet on listening to this passage, it appears that one piano is consistently playing the higher tone, with the other piano playing the lower tone.

The scale illusion can even be produced with inaccurate sound parameters. The video in the accompanying module shows two fifth graders at Atwater School, Wisconsin. They are playing the scale illusion on xylophones, with the class watching and listening, and it is evident that the classmates heard the effect clearly. Their music teacher, Walt Boyer, recorded this video.

**Scale Illusion on Xylophones**

http://dianadeutsch.net/ch02ex04

Let's examine the scale illusion more closely. As shown in Figure 2.5, it consists of a scale, with successive tones alternating from ear to ear. The scale is presented simultaneously in both ascending and descending form, such that when a tone from the ascending scale is in the right ear, a tone from the descending scale is in the left ear, and vice versa.

Just as with the octave illusion, it is quite rare for anyone to hear this pattern correctly. The type of illusion varies from one person to another, but the illusion most frequently obtained by right-handers consists of the correct sequence of pitches, but as two separate melodies—a higher one, and a lower one, that move in opposite directions. Furthermore, the higher tones all appear to be coming from the right earphone and the lower tones from the left one. When the earphone positions are reversed, this percept is maintained. So just as with the octave illusion, reversing the earphone positions creates the impression that the earphone that had been producing the higher tones is now producing the lower tones, and that the earphone that had been producing the lower tones is now producing the higher tones!

Right-handers and left-handers again differ statistically in terms of where the higher and lower tones appear to be coming from. Right-handers tend strongly to hear the higher tones on the right and the lower tones on the left, but left-handers as a group don't show this tendency. In addition, a different type of illusion is sometimes obtained: People hear only the higher tones, and little or nothing of the lower ones. Left-handers are more likely than right-handers to experience this type of illusion.

We can also note that the scale illusion, just as the octave illusion, involves illusory conjunctions. Take the listener who perceives this illusion with all the higher tones on the right and all the lower tones on the left (as shown in Figure 2.5). This listener is incorrectly conjoining the pitches of the higher tones that are presented to the left with the spatial location to the right, and incorrectly conjoining the pitches of the lower tones that are presented to the right with the spatial location

FIGURE 2.7. The pattern that produces the chromatic illusion, and a way it is often perceived. From Deutsch (1987).

to the left. The same goes for variants of the scale illusion that I created, and that we shall now explore.

One variant, which I named the *Chromatic Illusion*, is illustrated in Figure 2.7, and displayed in the "chromatic illusion" module. The pattern that produces this illusion consists of a two-octave chromatic scale that alternates from ear to ear in the same fashion as in the scale illusion. Most right-handers hear this pattern as a higher melodic line that travels down an octave and back, together with a lower melodic line that travels up an octave and back, with the two meeting in the middle. Yet in reality the tones produced by both the right channel and the left one are leaping around in pitch.[17]

Another related illusion that I created, called the *Cambiata Illusion*, is based on the same principle.[18] (I gave the illusion this name because a cambiata figure is a melodic pattern that circles around a central tone.) It can be heard either through stereo headphones or through stereo loudspeakers. The pattern that produces the illusion is illustrated on Figure 2.8 and portrayed in the "cambiata illusion" module. On listening to it through earphones, many people experience a repeating higher melody in one ear, and a repeating lower melody in the other ear. When the earphone positions are reversed, often the ear that had been hearing the higher

**SOUND PATTERN**

**PERCEPTION**

FIGURE 2.8. The pattern that produces the cambiata illusion, and a way it is often perceived. From Deutsch (2003).

melody continues to hear the higher melody, and the ear that had been hearing the lower melody continues to hear the lower melody. Yet in reality, the tones arriving at each ear are jumping up and down in pitch over a two-octave range.

Other people experience the cambiata illusion differently. For example, they might hear all the lower tones in their left ear, and only the higher tones separated by pauses in their right ear. As with the other stereo illusions we've been exploring, left-handers are more likely than right-handers to obtain complex and varied percepts.

Combinations of pitch and timbre can also give rise to illusory conjunctions. Michael Hall at the University of Nevada and his colleagues presented

**Cambiata Illusion**

http://dianadeutsch.net/ch02ex06

arrays of musical tones of varying timbres.[19] The tones were spatially distributed and presented simultaneously, and listeners searched for conjunctions of specific values of pitch and timbre. Illusory conjunctions were very common, occurring from 23 percent to 40 percent of the time. Such illusory conjunctions probably occur frequently in live orchestral music. For example, a listener might perceive a pitch as being performed by a brass instrument when in reality it's being produced by strings.

Illusory conjunctions also occur in memory. In one experiment, I presented listeners with a test tone, which was followed by four intervening tones, then by a pause, and then by a second test tone. The listeners were asked to ignore the intervening tones, and to judge whether the test tones were the same or different in pitch. On some occasions when the test tones differed, I included in the intervening sequence a tone of identical pitch to the second test tone. This produced a very large increase in the tendency of listeners to judge erroneously that the first and second test tones were identical in pitch. In other words, the listeners recognized correctly that a tone of the same pitch as the second test tone had occurred, but assumed falsely that it had been the first test tone.[20]

Tones that differ in pitch and duration can also produce illusory conjunctions in memory. William Thompson and his colleagues played listeners sequences of tones, each followed by a test tone. When the test tone matched the pitch of one of the tones in the sequence, and matched the duration of a different tone, the subjects frequently judged erroneously that the test tone had also occurred in the sequence.[21]

Illusory conjunctions must produce problems for conductors, since during a performance they should be able to identify where each note is coming from— so that, for example, they can target a performer who is playing a wrong note. It's possible that many conductors are less susceptible to illusory conjunctions than are most people. When I was visiting IRCAM (the Institute for Research and Coordination in Acoustics/Music in Paris), I had an interesting discussion with the late composer and conductor Pierre Boulez, who asked me to play him the octave and scale illusions. When he listened to the octave illusion through headphones, he was puzzled and disturbed by what he heard—my guess is that he was obtaining the illusion of a complex and changing pattern of sounds. Then I played him the scale illusion through loudspeakers. After listening intently for a while, he correctly described the patterns that were being played. Then I said to him, "So far you've been listening to the sounds coming from the loudspeakers analytically—one at a time. So now imagine that you are in front of an orchestra, and listen to both patterns as a whole instead." After a brief listen he suddenly said, "Now it's changed! I hear the right loudspeaker producing the higher tones and the left speaker producing the lower tones!" So Boulez was capable of listening to this complex pattern in different ways, and depending on his listening strategy he heard different melodic lines. This form of listening should be ideal for conductors, since they can then switch between evaluating the sounds produced by individual performers, and hearing the ensemble in its entirety.

Why should we experience the scale illusion, and other illusions like it? Our everyday world of sound is an orderly one, and our hearing mechanism, taking advantage of this orderliness, has developed assumptions about which sound patterns

FIGURE 2.9. Hermann Ludwig Ferdinand von Helmholtz.

are likely to occur. So when a pattern such as the scale illusion is played, our brain rejects the improbable, though correct, conclusion that two sources are each producing tones that jump around in pitch. Rather, it assumes that tones in one pitch range are coming from one source, and tones in a different pitch range are coming from a different source. So we perceptually reorganize the tones in space in accordance with this interpretation.

The view that our perception is strongly influenced by our knowledge and expectations was championed by the nineteenth-century German physicist Herman von Helmholtz (Figure 2.9), whose book *On the Sensations of Tone as a Physiological Basis for the Theory of Music* makes important reading even today.[22] The effect of unconscious inference (or top-down processing) on perception is apparent in the stereo illusions described here.

Another musical illusion that I created, in which you hear the correct sounds but hear them in the wrong locations, is called the *glissando illusion*.[23] It's produced by a synthesized oboe tone that's played together with a pure tone that glides up and down in pitch, rather like a siren. The two sounds switch repeatedly from side to side, such that whenever the oboe tone is on the right, a portion of the glissando is on the left, and vice versa.

**Glissando Illusion**

http://dianadeutsch.net/ch02ex07

To perceive the glissando illusion, sit in front of two loudspeakers, with one to your left and the other to your right. Then listen to it as presented in the Glissando Illusion module. People generally hear the oboe tone correctly as jumping back and forth between the loudspeakers; however, they hear the portions of the glissando as joined together quite seamlessly, so that it sounds like a single long tone. People localize the glissando in various ways. Right-handers are most likely to hear it traveling from left to right as its pitch moves from low to high, and then back from right to left as its pitch moves from high to low. But left-handers differ considerably among themselves in terms of the way in which the glissando appears to be traveling.

There's a further interesting twist to the way the glissando is perceived: It often appears to be traveling between low down in space (about level with the floor) when its pitch is at its lowest, and high up in space (about level with the ceiling) when its pitch is at its highest. Other researchers have noticed that listeners tend to hear higher tones as higher up in space, and lower tones as lower down.[24] In fact, the very words *high* and *low*, when applied to pitch, may reflect this tendency,

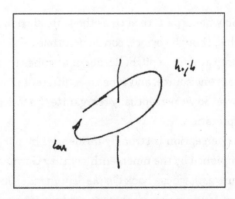

FIGURE 2.10. Diagram drawn by a listener to illustrate his perception of the glissando illusion. From Deutsch, Hamaoui, & Henthorn (2007).

though the reason why we hear sounds this way is mysterious. In the glissando illusion, the combination of apparent travel along the left–right dimension together with the low–high dimension often causes the listener to hear the glissando trace an illusory trajectory that moves diagonally across space, from low down on the left to high up on the right, and then back again. Figure 2.10 shows a drawing that was produced by a listener to illustrate his perception of the glissando illusion.

Now for all these illusions, right-handers tend to hear higher tones on their right and lower tones on their left, regardless of where they are really coming from. In addition, they tend to perceive combinations of tones more accurately

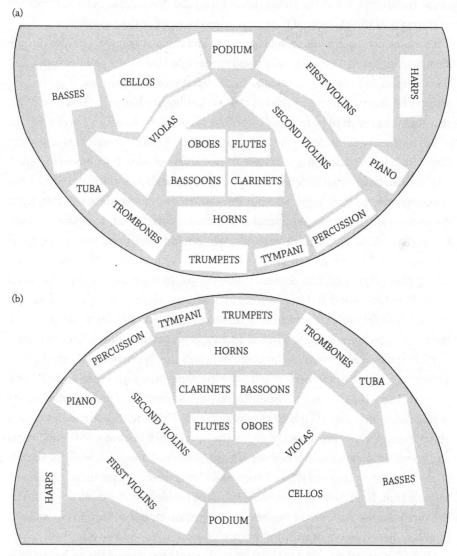

FIGURE 2.11. (a) A seating plan for the Chicago Symphony, as viewed from rear of stage. (b) The same arrangement as in (a), but as viewed from audience.

when the higher tones are to their right and the lower tones to their left.[25] To the extent that effects of this sort occur in listening to live music, this produces a conundrum. In general, contemporary seating arrangements for orchestras are such that, from the performer's point of view, instruments with higher registers are to the right, and those with lower registers are to the left. Figure 2.11(a) shows, as an example, a seating plan for the Chicago Symphony, viewed from the back of the stage. Let's consider the strings: The first violins are to the right of the second violins, which are to the right of the violas, which are to the right of the cellos, which are to the right of the basses. Then take the brasses: The trumpets are to the right of the trombones, which are to the right of the tuba. Notice also that the flutes are to the right of the oboes, and the clarinets to the right of the bassoons.[26]

The same principle tends to hold for choirs and other singing groups also: People singing in higher registers generally stand to the right (from their point of view), and those singing in lower registers stand to the left. Since it's important that the performers be able to hear themselves as well as possible, it's likely that this type of arrangement has evolved by trial and error, as it is conducive to optimal performance.

But here's a paradox: Since the audience sits facing the orchestra, this left–right disposition is, from their point of view, mirror-image reversed: As shown in Figure 2.11(b), instruments with higher registers are to the audience's left, and those with lower registers to their right. So from the audience's point of view, this left–right disposition should cause perceptual difficulties. In particular, instruments with low registers that are to the audience's right should tend to be poorly perceived and localized. There is a phenomenon known as "the mystery of the disappearing cellos" that plagues certain concert halls—people in the audience find the cellos difficult to hear—and architectural acousticians are not certain what causes this effect. No doubt there are a number of reasons, but it certainly doesn't help to have the cellos, which produce low-pitched sounds, positioned to the audience's right.

What can be done about this problem? We can't simply mirror-image reverse the orchestra, since then the performers wouldn't then hear each other so well. Let's suppose, then, that we turned the entire orchestra around 180 degrees, so that the performers would have their backs to the audience. This wouldn't provide a solution either, because the brasses and the percussion would then be closest to the audience, and so would drown out the delicate strings. Suppose, then, that we kept the orchestra with their backs to the audience, and also reversed the positions of the performers from front to back. This would solve the problem for the audience, but now the conductor would have the brasses and percussion closest to him, so he wouldn't be able to hear the strings well, and wouldn't be able to conduct so effectively.

One solution, that was jokingly suggested, would be to leave the orchestra as it is, but have the audience hanging upside down from the ceiling![27] This solution, however, would not be likely to be popular with concertgoers. On the other hand, when listening to music on your home stereo system, you might want to try reversing the left and right channels—this should produce greater perceptual clarity for most right-handers. There are some disadvantages to this scheme, however. In particular, the music would not sound the same as it does in concert halls, and so it might seem wrong in some way. As a further issue, as described by the composer Michael A. Levine, whereas orchestral performers strive for accuracy, the audience instead wants a holistic experience rather than one that has the greatest clarity. He reversed the classic layout in a recording, so that the higher instruments were to the listener's right and the lower ones to his left. He then listened as though naïvely, and considered the emotional and other overall effects of the music, and decided that the traditional arrangement sounded better. So great conductors and producers need to listen both ways—analytically, to evaluate the performances of the individual players, and holistically, as though they were naïve listeners in the audience. As I described earlier, Pierre Boulez was able to switch between analytic and synthetic forms of listening to the scale illusion.

At all events, taking these stereo illusions as examples, it's clear that we cannot assume that the music as it is heard is the same as the music that appears in the written score—or as it might be imagined on reading the score. There is a further complication to consider. As discussed earlier, although most people hear these illusions as with the higher tones to the right and the lower tones to the left, there are striking differences between listeners in how they are perceived. So not only do we hear these patterns "wrongly," but we also hear them differently from each other. These differences depend statistically on the listener's handedness, so that left-handers are more likely than right-handers to obtain varied illusions.

So far we have been focusing on passages in which two simultaneous streams of tones are played, one to the listener's right and the other to his left. The tones are perceptually reorganized in space, so that the melodies we hear are quite different from those that are really being presented. We now continue to explore peculiarities in hearing sequences of musical tones, this time involving further principles of perceptual organization.

*Note:* The octave illusion needs to be heard through stereo headphones or earbuds. The scale, chromatic, and cambiata illusions are best heard through stereo headphones or earbuds, but can also be heard through stereo loudspeakers. The glissando illusion is best heard through stereo loudspeakers.

# 3

## The Perceptual Organization of Streams of Sound

⌒

Music is organized sound.
—EDGARD VARÈSE

IMAGINE THAT YOU are sitting at a sidewalk cafe. Traffic rumbles past you along the street. Pedestrians stroll by, engaged in lively conversation. Waiters clatter dishes as food is served, and music emanates softly from inside the building. At times one of these streams of sound engages your attention, at times a different one, but in a sense you are continually aware of all the sounds that reach you.

Or imagine that you are listening to an orchestral performance in a concert hall. The sounds produced by the different instruments are mixed together and distorted in various ways as they travel to your ears. Somehow you are able to disentangle the components of the sound mixture that reaches you, so that you hear the first violins playing one set of tones, the flutes another, and the clarinets another. You group together the sounds that you perceptually reconstructed so as to hear melodies, harmonies, timbres, and so on. What algorithms does the auditory system employ to accomplish these difficult tasks?

Our hearing mechanism is constantly identifying and grouping sounds into unified streams, so that we can focus attention on one of them while others are relegated to the background. Then we can change our focus so that the foreground becomes background, and vice versa. Composers exploit this effect in a number of ways. In much accompanied singing, the voice is clearly intended as figure, and the accompaniment as background. On the other hand, in contrapuntal music, such as

FIGURE 3.1. Rubin vase.

many works by Johann Sebastian Bach, there are figure–ground relationships of a different type. Here two or more melodic lines are played in parallel, and the listener switches attention back and forth between these lines. In such music we have analogies to ambiguous figures in vision; these can be perceived in different ways depending on how the viewer focuses his attention. One example, created by the Danish artist Edgar Rubin, is shown in Figure 3.1.[1] This picture can be interpreted either as a white vase against a black background, or as two black faces staring at each other across a white space. We can choose at will to "see" either the vase or the faces, but we cannot achieve both perceptual organizations at the same time.

What perceptual principles do we employ when we group elements of an array into unified wholes? The Gestalt psychologists, who flourished during the early part of the twentieth century, were concerned with how the brain organizes the information presented to the senses so that we perceive entire figures, rather than collections of unrelated parts. Notable among the Gestaltists were Kurt Koffka, Max Wertheimer, and Wolfgang Kohler.[2] They proposed a set of grouping principles that have profoundly influenced the study of both hearing and vision even today (Figure 3.2).

One principle, which is termed *proximity*, states that we form connections between elements that are closer together in preference to those that are spaced further apart. In example (a) of Figure 3.2 we group the dots together on the basis of proximity, so we perceive them as grouped in pairs. A second principle, termed

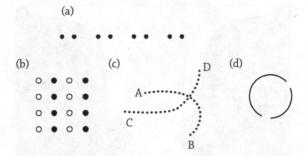

FIGURE 3.2. Illustrations of certain Gestalt principles of perceptual organization. These are (a) proximity, (b) similarity, (c) good continuation, (d) closure. From Deutsch (1980c).

*similarity*, states that we form connections between elements that are similar to each other in some way. In example (b), we group together the filled circles and separate them from the unfilled ones.

A third principle, termed *good continuation*, asserts that we form connections between elements that follow each other smoothly in the same direction. In example (c) we perceive lines AB and CD, rather than AC and DB. Yet another principle, termed *closure*, states that we tend to perceive elements of an array as complete units. In example (d) we perceive a circle with two gaps, rather than two separate curved lines. A further principle, termed *common fate*, maintains that elements that move in the same direction are perceptually linked together, Imagine a flock of birds all flying in the same direction. If one of these birds turns and flies in the opposite direction, it will stand out perceptually from the others, which we continue to see as a group.

As I contend in my article "Musical Illusions,"[3] and Albert Bregman explores in detail in his influential book *Auditory Scene Analysis*,[4] we can assume that the perceptual system has evolved to form groupings in accordance with these principles because they enable us to interpret our environment most effectively. This in turn provides an evolutionary advantage—early in our prehistory it would have enabled us to locate sources of food and alert us to the existence of predators.

Consider, for example, the principle of proximity. In vision, elements that are close together in space are more likely to have arisen from the same object than those that are further apart. In hearing, sounds that are close in pitch, or in time, are more likely to have come from the same source than are sounds that are far apart. The principle of similarity is analogous. Regions of the visual field that are similar in color, brightness, or texture have probably emanated from the same object, and sounds that are similar in character (a series of thuds, or chirps, for example) are likely to have come from the same source. It is probable that a line that follows a smooth pattern has originated from a single object, and a sound that changes smoothly in pitch has probably come from a single source. The same argument holds for common fate. An object that moves across the visual field gives rise to perceptual elements that move coherently with each other, and the components of a complex sound that rise and fall in synchrony are likely to have arisen from the same source.[5]

ఐ

While the visual system tends to carve the perceptual array into spatial regions that are defined according to various criteria such as brightness, color, and texture, the hearing system often forms groupings based on patterns of pitches in time. Such groupings can be illustrated by mapping pitch into one dimension of visual space and time into another. Indeed, musical scores have evolved by convention to be just such mappings. Tones that are higher in pitch are represented as higher on a musical staff, and progression in time is represented as movement from left to right. In the musical score shown in Figure 3.3, the dramatic visual contours reflect well the sweeping movements of pitch that are heard in this Mozart passage.

We have already seen an example of grouping by pitch proximity in the scale illusion, and in related stereo illusions. There, two series of tones arise in parallel from opposite regions of space, and each series is composed of tones that leap around in pitch. We reorganize the tones perceptually in accordance with pitch proximity, so that we hear one series of tones formed of the higher pitches as though coming from one region of space, and another series of tones formed of the lower pitches as though coming from the opposite region. So in these stereo illusions, organization by pitch proximity is so powerful as to cause half the tones to be perceived as coming from the incorrect location.

Other grouping effects appear strongly when sounds are presented in rapid succession. Musicians have known for centuries that when a series of tones is played at a rapid tempo, and these tones are drawn from two different pitch ranges, the series splits apart perceptually so that two melodic lines are heard in parallel. This effect employs the technique of *pseudopolyphony*, or *compound melodic line*. When we listen to such a pattern, we don't form perceptual relationships between tones

FIGURE 3.3. Excerpt from Mozart's Prelude and Fugue in C major.

that are adjacent in time—instead, we perceive two melodic lines in parallel, one corresponding to the higher tones and the other to the lower ones. Baroque composers such as Bach and Telemann employed this technique frequently. Figure 3.4 shows two examples. Parts (a) and (b) show the sequences in musical notation, and (a′) and (b′) show the same sequences plotted as pitch versus time. In passage (a) we perceive two melodic lines in parallel, one formed of the higher tones and the other of the lower ones. In passage (b) a single pitch occurs repeatedly in the lower range, and this provides the ground against which we perceive the melody in the higher range. Many examples of pseudopolyphony also occur in nineteenth- and twentieth-century guitar music, for example by Francisco Tárrega and by Augustín Barrios.

G. A. Miller and G. A. Heise at Harvard University created one of the first laboratory demonstrations showing that listeners organize rapidly repeating tones on the basis of pitch proximity. They had subjects listen to series of tones that consisted of alternating pitches played at a rate of ten per second.[6] When the alternating tones were less than three semitones apart, the listeners heard the series as a single coherent string (a trill). When, however, the pitch separation was more than three semitones, the listeners didn't form connections between tones that were adjacent in time, but instead perceived two repeating pitches that appeared unrelated to each other.

As a consequence of the perceptual splitting of pitch series into two streams, temporal relationships across tones in the different streams appear blurred. Albert Bregman and Jeffrey Campbell at McGill University created a repeating series of six pure tones, three high and three low ones, with those drawn from the high and low groups occurring in alternation. The rate of presentation was fast, and this caused the listeners to hear two distinct streams, one formed of the high tones and the other of the low ones. They could easily judge the order of the tones in the same stream, but were quite unable to judge their order across streams. Indeed, the listeners often reported that all the high tones preceded all the low ones, or vice versa, yet such orderings never occurred.[7]

**Galloping Rhythm**

http://dianadeutsch.net/ch03ex01

Even when the rate of presentation of such tones is slowed down so that we can identify their order, there is still a gradual breakdown in the ability to judge temporal relationships between tones as the pitch distance between them increases. The physicist Leon Van Noorden at Eindhoven, the Netherlands, created a repeating stream of tones in the continuing pattern . . .— ABA—ABA—. . . where the dashes indicate silent

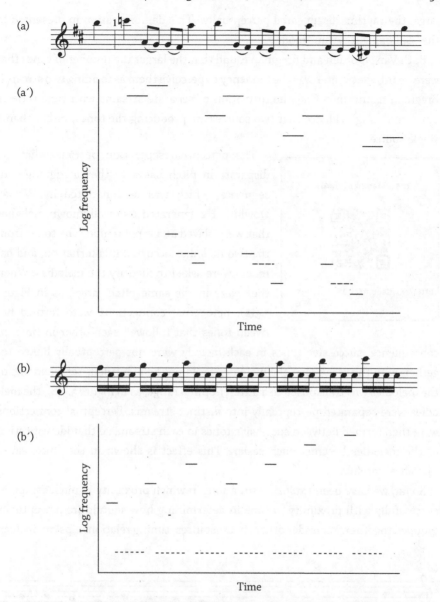

FIGURE 3.4. Grouping of melodic patterns by pitch proximity. In (a) we hear two parallel melodies, each in a different pitch range. In (b) a single pitch is repeatedly presented in the lower range, and this forms a ground against which the melody in the upper range is perceived. (a) from Telemann, Capriccio for recorder and basso continuo. (b) From Telemann, Sonata in C major for recorder and basso continuo. From Deutsch (1975a).

intervals. This stream was played with the tones beginning far apart in pitch, and their pitches gradually converged and then diverged. When the pitch difference between successive tones was small, listeners perceived the pattern as forming a "galloping rhythm" (di-di-dum, di-di-dum). But when this pitch difference was large, the rhythm disappeared perceptually.[8] This demonstration in presented in the "Galloping Rhythm" module.

Both Van Noorden and Bregman found that the larger the number of tones that were heard, the greater was the tendency to perceive them as forming two streams. Bregman maintained that the formation of separate streams with time reflects our cumulating evidence that two sources are producing the tones, rather than a single source.

**Interleaved Melodies**

http://dianadeutsch.net/ch03ex02

The perceptual separation of tones that are disparate in pitch has a further intriguing consequence, which was demonstrated by W. Jay Dowling. He generated two well-known melodies that were played at a rapid tempo. The tones from the two melodies occurred in alternation, and listeners were asked to identify the melodies. When they were in the same pitch range, as in Figure 3.5(a), perceptual connections were formed between tones that followed each other in time. In consequence, successive tones in each melody were not perceptually linked together, with the result that the melodies could not be identified. Yet when one of the melodies was transposed to a different pitch range, as in Figure 3.5(b), the melodies were separated perceptually into distinct streams. Perceptual connections were then formed between successive tones in each stream, so that identification of the melodies became much easier.[9] This effect is shown in the "Interleaved Melodies" module.

So far, we have been exploring situations in which proximity in pitch competes successfully with proximity in time in determining how sounds are perceptually grouped together. Yet under other circumstances, timing relationships can instead

FIGURE 3.5. Two well-known melodies played rapidly with the notes interleaved in time. In (a) the two melodies are in the same pitch range, and listeners find it difficult to separate them out and identify them. In (b) the melodies are in different pitch ranges, and so are easy to separate out and identify. From Dowling & Harwood (1986).

be the deciding factor. For example, if you take a sequence of sounds, and insert pauses between the sounds intermittently, listeners perceptually group them into units that are defined by the pauses. This tendency can be so strong as to interfere with grouping along other lines. Gordon Bower at Stanford University has shown that when a meaningful string of letters (acronyms or words) is read out to listeners, their memory for the strings worsens considerably when pauses are inserted that are inconsistent with their meanings. For example, we have difficulty remembering the string of letters IC BMP HDC IAFM. On the other hand, when the same sequence of letters is presented as ICBM PHD CIA FM, we perceive the acronyms, and so can easily remember the letter strings.[10]

I later showed that the same principle holds for musical passages. I devised passages that consisted of subsequences that had a uniform pitch structure. Musically trained participants listened to these passages, and wrote down in musical notation what they heard. When pauses were inserted between the subsequences so as to emphasize the repeating pitch structure, the participants notated the passages easily. On the other hand, when the pauses conflicted with the

**Timing and Sequence Perception**

http://dianadeutsch.net/ch03ex03

repeating pitch structure, they had considerable difficulty notating the passages. Figure 3.6 illustrates an example, which you can also hear in the "Timing and Sequence Perception" module. It's clear that when the pauses are in accordance

FIGURE 3.6. Three different types of temporal segmentation of a hierarchically structured passage. This consists of four subsequences that each contains three tones, with the second tone a semitone lower than the first, and the third tone a semitone higher than the second. (a) No temporal segmentation. (b) Temporal segmentation in accordance with the structure of the passage. (c) Temporal segmentation in conflict with the structure of the passage. From Deutsch (1999).

FIGURE 3.7. Passage from the beginning of the second movement of Beethoven's Spring Sonata for violin and piano. The tones played by the two instruments overlap in pitch; however, the listener perceives two melodic lines in parallel, corresponding to those played by each instrument. This reflects perceptual grouping by similarity. From Deutsch (1996).

with the repeating pitch structure (b), it is easy to apprehend the overall passage—however, inappropriate pauses (c) interfere with the ability to perceive the passage, despite its repeating pitch structure.[11]

We are amazingly sensitive to the quality of the sounds that we hear, and we readily group sounds into categories. This is an example of the principle of similarity. The English vocabulary contains an enormous number of words that describe different types of sound, with each word evoking a distinct image. Consider, for example, the abundance of words that describe brief sounds, such as *click, clap, crack, pop, knock, thump, crash, splash, plop, clink,* and *bang.* Then there are longer-lasting sounds such as *crackle, rustle, jangle, rumble, hum, gurgle, rattle, boom, creak, whir, shuffle,* and *clatter.* The richness of this vocabulary shows how much we rely on sound quality to understand our world.

**Passage from Beethoven's Spring Sonata**

http://dianadeutsch.net/ch03ex04

It's not surprising, then, that sound quality or timbre is a strong factor in determining how musical sounds are perceptually grouped together[12]. Composers frequently exploit this feature to enable the listener to separate out parallel streams in musical passages. In the passage shown in Figure 3.7, which is taken from Beethoven's Spring Sonata for violin and piano, the pitches produced by the two instruments overlap, yet we hear the phrases produced by the different instruments as quite distinct from each other. This can be heard in the "Beethoven's Spring Sonata" module.[12]

The psychologist David Wessel generated a series of three tones that repeatedly ascended in pitch, but tones that were adjacent in time were composed of alternating timbres. An example is illustrated in Figure 3.8, and displayed in the

FIGURE 3.8. Wessel's timbre illusion. A three-tone ascending pitch line is repeatedly presented, with successive tones alternating in timbre. When the difference in timbre is large, and the tempo of the passage is rapid, listeners perceive the pattern on the basis of timbre, and so hear two interwoven descending pitch lines. From Wessel (1979).

"Timbre Illusion" module. When the difference in timbre between the tones was small, listeners heard this passage as ascending pitch lines. Yet when this difference was large, listeners grouped the tones on the basis of timbre, and so heard two interwoven descending lines instead. Speeding up the tempo increased the impression of grouping by timbre.[13]

Timbre Illusion

http://dianadeutsch.net/ch03ex05

Just as with perceptual separation on the basis of pitch, separation by sound quality can have striking effects on our ability to identify the ordering of sounds. Richard Warren and his colleagues at the University of Wisconsin constructed a repeating series of four unrelated sounds—a high tone, a hiss, a low tone, and a buzz—which they presented at a rate of five sounds per second. Listeners were completely unable to name the order in which the sounds occurred—the duration of each sound had to be slowed down considerably before they were able to name their ordering.[14]

Much information arrives at our sense organs in fragmented form, and the perceptual system needs to infer continuities between the fragments, and to fill in the gaps appropriately. For example, we generally see branches of trees when they are partly hidden by foliage, and we infer which of the visible segments were derived from the same branch. When we make such inferences, we are employing the principles of good continuation and closure, since the gaps between segments of a branch can be filled in perceptually to produce a smooth contour.

The Kanizsa triangle shown in Figure 3.9 is an example of an illusory contour that is produced when the visual system fills in gaps so that an object is perceived as an integrated whole. We interpret this figure as a white triangle that occludes other objects, in accordance with the principles of good continuation and closure.[15]

As a related effect, our hearing mechanism is continually attempting to restore lost information. Imagine that you are talking with a friend on a busy street, and the passing traffic intermittently drowns out the sound of his voice. To follow

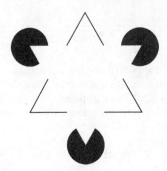

FIGURE 3.9. The Kanizsa triangle. We perceive a white triangle, though it is not in fact presented—rather, it is inferred based on the principles of closure and good continuation.

what your friend is saying, you need to infer continuity between the fragments of his speech that you can hear, and to fill in the fragments that you missed hearing. We have evolved mechanisms to perform these tasks, and we employ them so readily that it's easy to produce illusions based on them. In other words, we can easily be fooled into "hearing" sounds that are not really there.

One way to create continuity illusions is to have a softer sound intermittently replaced by a louder one; this creates the impression that the softer sound is continually present. George Miller and Joseph Licklider presented a tone in alternation with a louder noise, with each sound lasting a twentieth of a second. Listeners reported that the tone appeared to continue right through the noise.[16] The authors obtained similar effects using lists of words instead of tones, and they described such illusions as like viewing a landscape through a picket fence, where the pickets interrupt the view at intervals, and the landscape appears continuous behind the pickets.

Continuity effects can also be produced by signals that vary in time. Gary Dannenbring at McGill University created a gliding tone that alternately rose and fell in pitch. He then interrupted the glide periodically. When only silence intervened between portions of the glide, as in Figure 3.10(a), listeners heard a

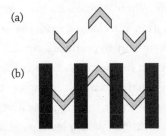

FIGURE 3.10. Illustration of the continuity effect using gliding tones. When the gliding tone is interrupted by silent gaps (a), it is heard as fragmented. But when loud noise bursts are inserted during the gaps (b), the glide appears to continue through the noise. The gray lines represent the gliding tone, and the black bars represent the noise. From Bregman (1990).

sequence of isolated V-shaped patterns. Yet when loud noise bursts were inserted during each gap, as in Figure 3.10(b), they instead heard a single, continuously rising and falling sound that appeared to glide back and forth right through the noise.[17]

A particularly convincing version of the continuity illusion was created by Yoshitaka Nakajima in Japan, and can be heard in the "Continuity Illusion" module. First there is a long tone with a gap in the middle, and the gap is clearly heard. Then a loud complex tone is presented alone. Finally, the long tone with the gap is again presented, this time with the complex tone inserted in the gap, and the long tone now appears continuous.

**Continuity Illusion**

http://dianadeutsch.net/ch03ex06

Speech sounds are also subject to illusory restoration effects, and these are particularly strong when a meaningful sentence establishes a context. Richard Warren and his colleagues recorded the sentence "The state governors met with their respective legislators in the capital city." Then they deleted the first "s" in the word "legislators" and replaced it with a louder sound, such as a cough, and found that the sentence still appeared intact to listeners. Even after hearing the altered sentence several times, the listeners didn't perceive any part of it as missing, and they believed that the cough had been added to the recording. Even more remarkably, when they were told that the cough had replaced some portion of the speech, they couldn't locate where in the speech it had occurred. Yet when the missing speech sound left a silent gap instead, the listeners had no difficulty in determining the position of the gap.[18]

Similar effects occur in music. Takayuki Sasaki in Japan recorded familiar piano pieces, and replaced some of the tones with bursts of noise. Listeners heard the altered recordings as though the noise bursts had simply been added to them, and they generally couldn't identify the positions of the noise bursts.[19] Perceptual restorations of this type must occur frequently when listening to music in concert halls, where coughs and other loud noises produced by the audience would otherwise cause the music to appear fragmented.

The continuity illusion is used to good effect in Electronic Dance Music (EDM). The beat from the kick drum is sometimes accompanied by a subtle reduction in amplitude of other ongoing sounds. As a result of the continuity effect, the kick can be heard more clearly, while the other sounds still appear continuous.

What characteristics of the substituted sound are conducive to producing illusory continuity? When a sound signal is interrupted by periods of silence, it's unlikely that an extraneous factor has caused these interruptions, so we can reasonably assume that the interruption is occurring in the signal itself. But loud noise bursts that replace the silent gaps can plausibly be interpreted as extraneous sounds that are intermittently drowning out the signal. For this reason, illusory continuity effects occur best when the interposed sound is loud, when it is strictly juxtaposed with the signal, and when the transition between it and the signal is sufficiently abrupt as to avoid the conclusion that the signal itself is intermittent.

However, these conditions are not necessary for continuity effects to occur. For example, guitar tones are characterized by rapid increases and decreases in loudness. In music played on such instruments, when the same tone is rapidly repeated many times, and it's intermittently omitted and replaced by a different tone, listeners generate the missing tone perceptually. Many such examples occur in guitar music of the nineteenth and twentieth centuries, such as Tárrega's *Recuerdos de la Alhambra* (see Figure 3.11), and Barrios's *Una Limosna por el Amor de Dios*. Here the strong expectations set up by the rapidly repeating tones (called 'tremolo') cause the listener to "hear" the missing tones, even though they are not being played. The "Recuerdos de la Alhambra" module presents such an example.

**Passage from Tárrega's Recuerdos de la Alhambra**

http://dianadeutsch.net/ch03ex07

Another approach to the perception of musical streams exploits the use of orchestral sound textures to create ambiguous images. In the late nineteenth and early twentieth centuries, composers such as Debussy, Strauss, and Berlioz used this approach to remarkable effect. Later composers such as Varèse, Penderecki, and Ligeti made sound textures the central elements of their compositions. Rather than following the principles of conventional tonal music, they focused on sound masses, clouds, and other textural effects, with extraordinary results. Ligeti's *Atmospheres* is a fine example; it was made particularly famous by the mysterious ambience it created in Stanley Kubrick's movie *2001: A Space Odyssey*. As Philip Ball wrote:

> The main effect of this harmonic texture is to weave all the voices together into one block of shifting sound, a rumbling, reverberant, mesmerizing mass. Faced with such acoustic complexity, the best the mind can do is to lump it all together, creating a single "object" that, in Bregman's well-chosen words, is "perceptually precarious"—and therefore interesting.[20]

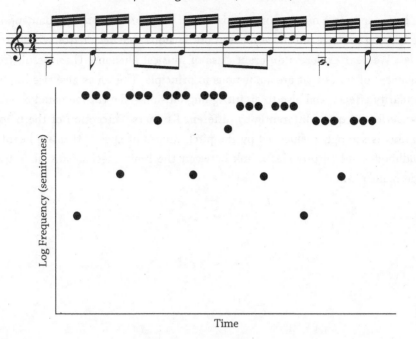

FIGURE 3.11. The beginning of *Recuerdos de la Alhambra*, by Tárrega. Although the tones are played one at a time, we perceive two parallel lines in accordance with pitch proximity, and we fill in the missing tones in the repeating sequence. From Deutsch (1996).

Such sound objects can appear to be in a perpetual state of transformation. In this way they are like visual arrays that seem to be constantly changing. Faced with ambiguity, the perceptual system is constantly forming organizations and then replacing them with new ones. Most of us have gazed into the sky and seen changing images in the clouds—strange edifices, faces, figures, hills, and valleys. These perceptual impressions have fascinated poets and artists, who sometimes deliberately incorporated images of objects in paintings of clouds or landscapes. As described by Shakespeare in *Antony and Cleopatra*:

> Sometime we see a cloud that's dragonish,
> A vapour sometime like a bear or lion,
> A tower'd citadel, a pendant rock,
> A forked mountain, or blue promontory
> With trees upon't, that nod unto the world,
> And mock our eyes with air.[21]

We have explored a number of organizational principles that are fundamental to the hearing process, and have focused on illusions that result from these principles. We next explore a different class of musical illusions. These result from sequences of tones that are ambiguous in principle. The tones give rise to pitch circularity effects, and also to an intriguing illusion—the tritone paradox—that is experienced quite differently by different listeners. Perception of the tritone paradox is strongly influenced by the pitch ranges of speech that are heard in childhood—and so provides a link between the brain mechanisms underlying speech and music.

# 4

## Strange Loops and Circular Tones

"Escher for the Ear"
—P. YAM, *Scientific American*, 1996[1]

EPIMENIDES, A CRETAN philosopher who lived around 600 B.C., made the famous pronouncement "All Cretans are liars," so creating a paradox that has been discussed by logicians ever since. It goes like this: As Epimenides was himself a Cretan, it follows from this statement that he was a liar. But if he was a liar, then his statement was untrue—so Cretans are not liars, but tell the truth. But if Cretans tell the truth, then the statement "All Cretans are liars" must be true. But then. . . . And we can go on like this forever.

It has been pointed out that the Epimenides Paradox is not watertight, a better one being the Liar Paradox, which refers to the statement "I am lying", meaning that the statement being made is untrue. But if it is untrue then it is a lie, so the statement "I am lying" is true. But if the statement is true, then I am not lying, so the statement is false—and so we are made to run around in circles indefinitely.

In his book *Gödel, Escher, Bach,*[2] Douglas Hofstadter coined the term *Strange Loop* to refer to a phenomenon that occurs whenever, in moving through the levels of a hierarchical system, we find ourselves back where we started. He gave the Liar Paradox as a stark example of a loop that can be traversed infinitely.

The Dutch artist M. C. Escher used Strange Loops in many of his works. Perhaps his best known is the lithograph *Ascending and Descending*, shown in Figure 4.1.[3] Here we see monks trudging up and down a staircase in endless journeys.

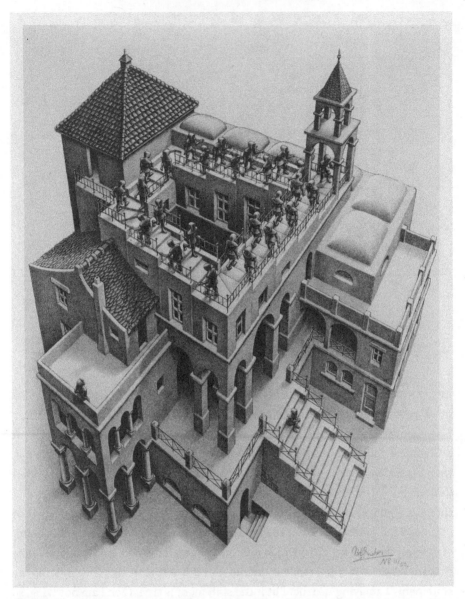

FIGURE 4.1. M. C. Escher's *Ascending and Descending*. © 2018 The M. C. Escher Company—The Netherlands. All rights reserved. www.mcescher.com.

This lithograph was based on the "impossible staircase" devised by Penrose and Penrose,[4] à variant of which is shown in Figure 4.2. Each step that's one step clockwise from its neighbor is also one step downward, so the staircase seems to descend (or ascend) endlessly. Our perceptual system insists on this interpretation, even though we know that it must be incorrect. In principle, we could view the figure correctly as depicting four sets of stairs that are discontinuous and seen

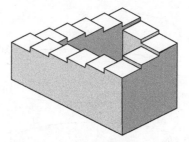

FIGURE 4.2. An impossible staircase, From Penrose & Penrose (1958).

from a unique perspective—yet we never perceive it this way. Our perceptual system opts for the simplest interpretation, even though it is nonsensical.

In the early 1960s, the psychologist Roger Shepard produced a series of tones that was analogous to the Penrose staircase (Figure 4.2).[5] By analogy with real-world staircases, pitch is often viewed as extending along a one-dimensional continuum from low to high. This is known as *pitch height*, and we can experience it by sweeping our hand from left to right up a piano keyboard. But pitch also has a circular dimension known as *pitch class*, which defines the position of a tone within the octave. The system of notation for the Western musical scale acknowledges the circular dimension by designating each note both by a letter name, referring to the position of the tone within the octave, and also by a number, referring to the octave in which it is placed. So as we ascend the scale in semitone steps, we repeatedly traverse the pitch class circle shown in Figure 4.3 in clockwise direction, so that we play C, C♯, D, D♯, and so on around the circle, until we reach C again, but now the note is an octave higher.

In order to accommodate both the linear and the circular dimensions of pitch, a number of theorists, going back at least to the nineteenth-century German mathematician Moritz Drobisch, have proposed that pitch should be represented as a

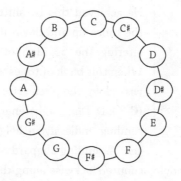

FIGURE 4.3. The pitch class circle.

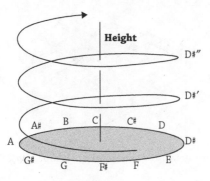

FIGURE 4.4. A pitch helix. From Shepard (1965).

helix that has one complete turn per octave, so that pairs of tones that are separated by octaves are close together in this helical space[6] (see Figure 4.4). Shepard reasoned that it might be possible to minimize the dimension of height in this space, while preserving the dimension of pitch class, so that all tones standing in octave relation would be mapped onto a single tone, which would have a well-defined pitch class but an indeterminate height. So the helix would be collapsed into a circle, and judgments of relative pitch would be completely circular.

Shepard devised a brilliant algorithm to achieve this goal. He synthesized a bank of complex tones, each of which consisted of ten frequency components (pure tones) that were separated by octaves. A fixed, bell-shaped spectral envelope determined the amplitudes of the components. Those in the middle of the musical range were highest in amplitude, and the others gradually decreased in amplitude along either side of the log frequency continuum. Since all the frequency components were related by octaves, the pitch class of each complex tone was well defined perceptually; however, the other components that were needed to define the tone's perceived height were missing[7].

**Ascending Shepard Scale**

http://dianadeutsch.net/ch04ex01

Shepard then varied the pitch classes of the tones, while their pitch heights remained fixed. He achieved this by shifting the components of the tones up and down in log frequency without altering the position of the spectral envelope. When this bank of tones was played moving clockwise along the pitch class circle in semitone steps (C, C#, D, D#, . . .), listeners heard the series as ascending endlessly in height—as can be heard in the "Ascending Shepard Scale" module. And when the tones were played moving counterclockwise along the circle (C, B, A#, A, . . .) they appeared to descend endlessly. Because this effect is so intriguing, it has been generated to accompany numerous videos of bouncing balls, stick figures, and other objects traversing the Penrose staircase, with each step accompanied by a step along the Shepard scale.

Circularity effects produced by Shepard tones also depend on the Gestalt principle of proximity, which was described in Chapter 3. Let's suppose that the pattern C♯–D is played using Shepard tones. In linking them together, the perceptual system has three options. First, since at the physical level the circle is flat with respect to height, the tones might be perceived as identical in pitch height, so that neither an ascending nor a descending pattern would be perceived. Second, the perceptual system might follow the principle of proximity, and link the tones together so that the shorter distance between them along the circle would be traversed; in this case C♯–D would be heard as ascending (Figure 4.5a). Third, the perceptual system might opt to traverse the longer distance between the tones, in which case C♯–D would be heard as descending instead (Figure 4.5b).

It turns out that for a semitone separation, the perceptual system always links Shepard tones together on the basis of proximity. A tone pair that moves one semitone clockwise along the circle (such as D–D♯, or E–F) is always heard as ascending, and a tone pair that moves one semitone counterclockwise (such as D♯–D or F–E) is always heard as descending. Circular scales can therefore be produced using such a bank of tones: C–C♯ is heard as ascending, as is C♯–D, D–D♯, D♯–E, and so on around the circle, so that A–A♯ is also heard as ascending, as is A♯–B, and B–C. The scale therefore ends in the same place as it begins. When a longer distance between the tones along the circle is traversed, such as C–E, or D♯–F♯, the tendency to link the tones together based on proximity persists, but is somewhat diminished.

The proximity principle is also involved in viewing the Penrose staircase. Notice that as we move from each step to its clockwise neighbor, we move one step downward. Because this is true of all the relationships between each step and its clockwise neighbor, the impression is produced of a staircase that descends (or ascends)

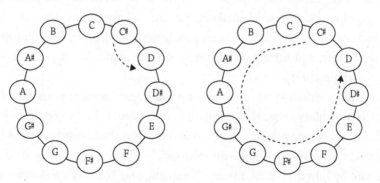

FIGURE 4.5. Two different ways in which Shepard tones could in principle be linked together.

endlessly. As Shepard pointed out, proximity can create a similar illusion of visual motion. Suppose you set up a row of lights extending from left to right, with every tenth light turned on. Then suppose you turned off all these lights simultaneously, and at the same time turned on all the lights just to the right of them. Because the visual system invokes proximity, we perceive the entire row of lights as shifting one position to the right. In principle, we could instead perceive the lights as shifting nine positions to the left, but this never happens.

**Descending Risset Glide**

http://dianadeutsch.net/ch04ex02

The French composer and physicist Jean-Claude Risset generated a powerful variant of Shepard's effect. Rather than a set of tones that moved stepwise along the pitch class circle, he created a single tone that glided around it. When the glide was in a clockwise direction, listeners perceived the tone as gliding endlessly up in pitch. When it moved in a counterclockwise direction, listeners heard an endlessly descending glide—as you can hear in the "Descending Risset Glide" module.[8]

Risset employed such a descending glide in his incidental music to Pierre Halet's play *Little Boy*, which depicts the nightmare of a pilot who was involved in the destruction of Hiroshima. Listening to this sound produces the impression of falling into a bottomless pit, and can be an emotional experience, appropriate to this tragic event. Risset also produced glides that appeared to be moving both upward and downward at the same time. He achieved this feat by having the components of the tones move upward while at the same time the spectral envelope moved downward, or vice versa.

Circular scales based on Shepard's algorithm can even be produced on the piano—though as we have only two hands, and these scales require playing tones that are separated by octaves, it helps to have two people playing together. However, it is possible for a single performer to achieve this with practice. At a gathering of magicians, puzzle makers, recreational mathematicians, and artists, I watched Scott Kim, the extraordinary puzzle designer, play an ascending Shepard scale repeatedly, and utterly convincingly, on the piano. Try it yourself—but it takes a lot of practice![9]

Since pitch proximity plays an important role in producing impressions of circularity, the possibility arises that circularity effects could also be generated by tone complexes whose components stand in ratios other than an octave. And indeed such effects have been created—for example, by Risset in a number of compositions, and by Edward Burns, Ryunen Teranishi, and Yoshitaka Nakajima in psychoacoustic experiments.[10]

Although pure pitch circularities were first generated in the mid-twentieth century when exact control of sound parameters became possible, composers have long been intrigued by the idea of circular pitch patterns, and have created musical arrangements approaching pitch circularity for centuries.[11] English keyboard music of the sixteenth and seventeenth centuries—for example, by the composer Orlando Gibbons—included ingenious manipulations of tone sequences that gave rise to circular effects. In the eighteenth century, Johann Sebastian Bach devised passages that produced circular impressions—an excellent example occurs in his Prelude and Fugue in E minor for organ, BWV 548.

In the first part of the twentieth century, Alban Berg used pitch circularity in his opera *Wozzeck*. He generated a continuously rising scale such that the higher instruments faded out at the high end of their range, while the lower instruments faded in at the low end. Composers such as Bartók, Prokofiev, and Ligeti also experimented with pitch circularity. Most notably, Jean-Claude Risset made extensive use of circular configurations in his orchestral works; for example, in his composition *Phases*.

Twentieth-century composers of electroacoustic music also employed pitch circularities. These effects occur, for example, in Risset's *Little Boy*, John Tenny's *For Ann (rising)*, and Karlheinz Stockhausen's *Hymnen*. Because they can be very dramatic, such effects have been used in movies. Richard King, the sound designer for the movie *The Dark Knight*, employed an endlessly ascending glide for the sound of Batman's vehicle, the Batpod. As he explained in the *Los Angeles Times*, "When played on a keyboard, it gives the illusion of greater and greater speed; the pod appears unstoppable."[12] For Christopher Nolan's 2017 movie *Dunkirk*, the composer Hans Zimmer created sound sequences inspired by Shepard scales, that gave an impression of ever-increasing tension.[13]

Eternally ascending and descending scales often have emotional effects. Many people have told me that listening to an eternally descending scale makes them feel subdued, and somewhat depressed. Once when I was playing such a scale to a large class, I found that, after about ten seconds of listening, many of the students were nodding their heads increasingly downward. The contrary happened when I played them an eternally ascending scale—it made the students perk up, and many of them declared that they felt energized.

Circularities of timing have also been produced. Jean-Claude Risset generated an eternally accelerating sequence using parallel streams of beats at different amplitudes, and whose tempi stood in the relationship of 2:1. He set the amplitudes of the streams such that those at the center of the range of tempo were always highest in amplitude, with the amplitudes of the other streams gradually decreasing as their tempi diverged from the center. As the sequence as a whole accelerated, the listener heard an accelerating pattern—but since the fastest stream faded out and the slowest stream faded in, it was heard to continue indefinitely.

∽

Returning to pitch circularity, we can ask: Is it necessary that our musical material be confined to highly artificial tones, or to several voices sounding simultaneously? Or might it be possible to create circular scales from single tones that are derived from natural instrument sounds? The ability to do this would greatly expand the scope of musical materials available to composers.

Tones that are produced by most natural instruments consist of many pure tones called *harmonics* (or overtones). The lowest harmonic, termed the *fundamental frequency*, corresponds to the perceived pitch, and the frequencies of the other harmonics are whole-number multiples of the fundamental. So if we call the fundamental frequency 1, the first six harmonics of a harmonic complex tone can be called 1, 2, 3, 4, 5, and 6. Then if we delete the odd-numbered harmonics (1, 3, and 5), we are left with the even-numbered harmonics (2, 4, and 6). Now, doubling the fundamental frequency of a harmonic complex tone causes it to move up an octave, so deleting the odd-numbered harmonics causes its pitch to be perceived an octave higher.

The physicist Arthur Benade has pointed out that a good flautist, while playing a steady note, can alter his manner of blowing in such a way as to vary the amplitudes of the odd harmonics relative to the even ones, and so produce an intriguing effect. Suppose the flautist begins with Concert A (a note with fundamental frequency 440 Hz); the listener hears this note as well defined both in pitch class and in height. Suppose, then, that the flautist gradually reduces the amplitudes of the odd harmonics relative to the even ones. At some point the listener realizes that he is no longer hearing Concert A but is instead hearing the A an octave higher. Yet the perceptual transition from the low A to the high one appears quite smooth.[14]

This observation implies that a tone consisting of a full harmonic series might be made to vary continuously in height without traversing the path specified by the helical model, and instead bypassing the helix and moving upward or downward between tones of the same pitch class—for example between D♯, D♯′ and D♯″ in the helix shown on Figure 4.4.

To try this, I began by presenting a tone at Concert A, consisting of six harmonics at equal amplitude. I then smoothly glided *down* the amplitudes of the *odd-numbered harmonics* until they could no longer be heard. And the predicted effect occurred—the tone appeared to glide slowly *up* in height, ending up sounding one octave above Concert A!

Based on this demonstration, I surmised that it might be possible to generate a circular bank of tones by systematically varying the relationships between the odd and even harmonics, while at the same time varying their pitch classes. So I began with a bank of twelve tones, each of which comprised the first six components of a harmonic series. The fundamental frequencies of the twelve tones varied in semitone steps over an octave's range. For the tone with the highest fundamental

**Circularity Illusions by Varying Harmonics**

http://dianadeutsch.net/ch04ex03

frequency, the odd and even harmonics were equal in amplitude. Then for the tone a semitone lower, I reduced the odd harmonics relative to the even ones, and this slightly raised the perceived height of the tone. Then for the tone another semitone lower, I further reduced the amplitude of the odd harmonics, so raising the perceived height of this tone to a greater extent. I continued moving down the octave in semitone steps, reducing the odd harmonics further with each step, until for the tone with the lowest fundamental, the odd harmonics no longer contributed to the perceived height of the tone. This tone was therefore heard as displaced up an octave, and pitch circularity was achieved. You can hear this effect in the "Circularity Illusions by Varying Harmonics" module.

In a formal experiment to determine whether such a bank of tones would indeed be perceived as circular, I created two such banks together with my colleagues Trevor Henthorn and Kevin Dooley. For each bank, each tone was paired with every other tone, both as the first tone of the pair and also as the second tone. Listeners judged whether each pair of tones ascended or descended in pitch. We found that judgments of these tone pairs were overwhelmingly determined by proximity along the pitch class circle. So, for example, the tone pair F–F♯ was always heard as ascending, and the pair E–D♯ was always heard as descending. When the tones forming a pair were separated by larger distances along the pitch class circle, the tendency to follow by proximity gradually lessened. And when we subjected our data to multidimensional scaling, we obtained circular configurations for both banks of tones.

As expected from these results, when such a bank of tones was played moving clockwise along the circle in semitone steps, a scale was heard that appeared to ascend endlessly. And when the tones were played moving counterclockwise in semitone steps, an endlessly descending scale was heard. Also using these parameters, we created long gliding tones that appeared to ascend or descend endlessly in pitch.[15] Since in this experiment circular scales were created from full harmonic series, the experiment has opened the door to transforming banks of natural instrument tones so that they would exhibit pitch circularity. William Brent (then a graduate student at the UCSD music department) used this algorithm to create endlessly ascending and descending scales from banks of bassoon tones. He also achieved some success with violin, flute, and oboe samples. This is still a work in progress, but it is exciting to envisage music that sounds as though played by natural instruments, yet in which the perceived heights of tones can be specified precisely, rather than being limited to octave intervals. In other words, rather than envisaging pitch as a helix, it can now be conceived as a solid cylinder.[16]

ᴄ᷄

We have been exploring some intriguing perceptual effects involving tones that have clearly defined pitch classes but are ambiguous with respect to height. When two such tones are played in succession, and these are separated by a short distance along the pitch class circle, our perceptual system invokes the principle of proximity, so that we hear a succession of such tones as circular. We now explore what happens when two such tones are presented when they are related by exactly a half-octave (a tritone) so proximity along the pitch class circle cannot be invoked. Will listeners' judgments then be ambiguous, or will the perceptual system invoke a further principle to resolve the ambiguity?

# 5

## The Tritone Paradox

### AN INFLUENCE OF SPEECH ON HOW MUSIC IS PERCEIVED

SO FAR WE have seen that when two tones are played in succession, and these are well defined in terms of pitch class, but ambiguous in terms of height, listeners hear an ascending or a descending pattern depending on which is the shorter distance between the tones along the pitch class circle. But we can then ask: What happens when a pair of such tones is presented that are related by a half-octave (an interval known as a tritone,[1]), so that the same distance between them would be traversed in either direction? What happens, for example, when you play C followed by F♯, or G♯ followed by D, and so on? When listeners are asked whether such a tone pair forms an ascending or descending pattern, will their judgments be ambiguous, or will the perceptual system invoke some other principle so as to avoid ambiguity?

In considering this question, it occurred to me that another cue is available to draw on: The listener could refer to the absolute positions of the tones along the pitch class circle. If tones in one region of the circle were perceptually tagged as higher, and those in the opposite region were tagged as lower, the ambiguity of height would be resolved. Rephrasing this hypothesis in detail, the listener could establish absolute positions for the tones along the pitch class circle. Let's envisage the circle as a clock face, and the pitch classes as numbers on the clock. Suppose a listener oriented the circle so that the tone C stood in the 12 o'clock position, C♯ in the 1 o'clock position, and so on, as in Figure 5.1. The listener would then hear the tone pair C–F♯ (and C♯–G and B–F) as descending, and the tone pair F♯–C (and

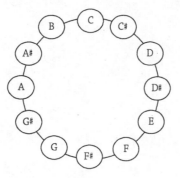

FIGURE 5.1. The pitch class circle.

G–C♯ and F–B) as ascending. Or if the listener oriented the circle so that G stood in the 12 o'clock position, he would hear the tone pair C♯–G (and C–F♯ and D–G♯) as ascending, and the tone pair G–C♯ (and F♯–C, and G♯ –D) as descending.

To explore this possibility, I played listeners pairs of Shepard tones, with the tones forming each pair related by a tritone.[2] Listeners judged for each pair of tones whether it ascended or descended in pitch. The hypothesis was strikingly confirmed—the judgments of most listeners showed systematic relationships to the positions of the tones along the pitch class circle, so that tones in one region of the circle were perceived as higher and tones in the opposite region were perceived as lower.

**Tritone Paradox**

http://dianadeutsch.net/ch05ex01

In addition, an entirely unexpected finding emerged: The orientation of the pitch class circle with respect to height differed completely from one listener to another.[3] For example, when presented with the tone pair D–G♯, some listeners clearly heard an ascending pattern, while other listeners clearly heard a descending one. Figure 5.2 shows the judgments of the tritone paradox made by two different listeners. Each plot shows, for one listener, the percentage of times that the pattern was heard as descending as a function of the pitch class of the first tone of the pair. For the listener whose plot is shown on the upper left, the 12 o'clock position of his pitch class circle was around C and C♯. Yet for the listener whose plot is shown on the lower left, the 12 o'clock position of his pitch class circle was around F♯ and G. I named these pairs of tones, together with their strange properties, the *tritone paradox*.

The best way to experience this strange phenomenon is to find a group of friends, and have them listen to the tone pairs presented in the "Tritone Paradox"

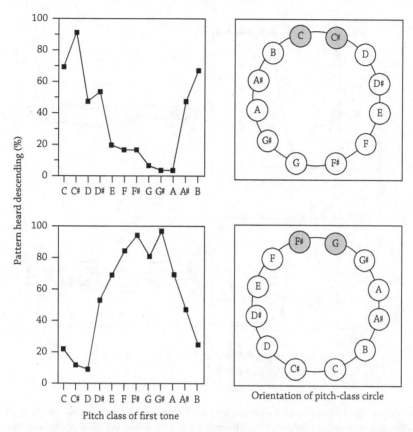

Pitch class of first tone

FIGURE 5.2. Judgments of the tritone paradox by two different listeners who heard the pattern in roughly opposite ways. The gray circles on the right are peak pitch classes, which are derived from the plots shown on the left. So the listener whose judgments are shown on the upper part of figure had peak pitch classes C and C♯, and the listener whose judgments are shown on the lower part of the figure had peak pitch classes F♯ and G. From Deutsch (1997).

module. Play them the first pair of tones, and ask them to judge whether they hear an ascending or a descending pattern. Then proceed to the next pair of tones, and ask them to make the same judgment. Continue until they have judged all four pairs of tones. I can pretty much guarantee that your friends, to their astonishment, will disagree among themselves as to which tone pairs they hear as ascending and which as descending. The best way to present this demonstration is in an auditorium with over a hundred people in the audience, and to ask for a show of hands for each judgment—in this way, the members of the audience will notice the strong perceptual disagreements among them.

These variations in perception of the tritone paradox have an intriguing consequence: Patterns composed of such tone pairs should be perceived by a group of listeners in many different ways. For example, a listener whose judgments are

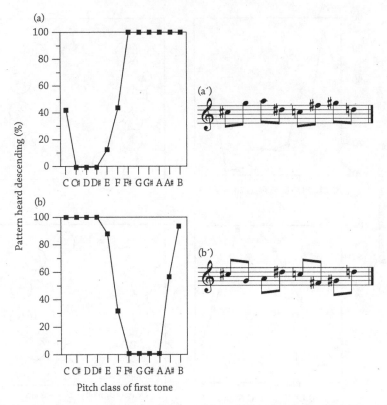

FIGURE 5.3. (a) Judgments of the tritone paradox by two listeners who heard the tritone paradox in a very orderly fashion, but differently from each other. (b) The ways in which these two listeners heard the identical eight-note pattern composed of tritones. From Deutsch (1999).

shown on Figure 5.3(a) would perceive the series of tone pairs C#–G / A–D# / C–F# / G#–D as in Figure 5.3(a′). However, a listener whose judgments of the same pattern are shown in Figure 5.3(b) would perceive the identical series of tone pairs as in Figure 5.3(b′) instead. So using this algorithm, we can compose melodies that are heard quite differently by different listeners!

A further implication of the tritone paradox concerns transposition. It's generally accepted that when a melody is played in one key, and it's then shifted to a different key, it's heard as the same melody (see Figure 2 of the Appendix). This is considered one of the cornerstones of music theory. But the tritone paradox violates this principle—these tritones are clearly heard as ascending when they involve one pair of tones, but as descending when they involve a different pair. As a result, an extended passage composed of these tritones will be perceived as one melody in one key, but as an entirely different melody when it's transposed to a different key!

So what can be the reason for the strange relationship between pitch class and perceived height that's revealed in the tritone paradox, and for the differences between listeners in how this pattern is perceived? I searched for correlates with musical training, but found none. I also found no relationships with simple characteristics of the hearing mechanism; I checked using audiometric tests, I presented the pattern at different overall amplitudes, and I compared judgments when the pattern was played to the left or right ears, but found no effect. I noticed, though, that the way I heard the tritone paradox tended to be roughly opposite from the way most of my students heard it. At the time—this was the 1980s—most of my students were English-speaking Californians, whose parents were also English-speaking Californians. Since I had grown up in London—in the south of England— it seemed to me, therefore, that speech patterns, particularly the pitch range of speech to which people were exposed in childhood, might be responsible for these perceptual differences. This was a long shot, but I also noticed that visitors from London tended to hear the tritone paradox similarly to the way I heard it.

More specifically, I surmised that each person has a mental representation of the pitch class circle, which is oriented in a particular direction with respect to height. The way the circle is oriented is derived from the speech patterns to which he or she has been most frequently exposed, particularly in childhood. This mental representation (or template) then determines both the pitch range of the person's speaking voice, and also how he or she hears the tritone paradox. And in a further experiment, my colleagues and I indeed found a correlation between the way listeners heard the tritone paradox and the pitch ranges of their speaking voices.[4]

Since the pitch ranges of speech vary with dialect, I carried out an experiment to explore whether perception of the tritone paradox also varies with the listener's dialect. I compared two groups. The first group consisted of speakers of English who had been born and grown up in California. The second group also consisted of speakers of English, but they had been born and grown up in the south of England. A striking difference between the two groups emerged: In general, when a listener from California heard a pattern as ascending, a listener from the south of England heard it as descending; and vice versa. I plotted the data for each listener as a function of the peak region of his or her pitch class circle, as derived from the way he or she heard the tritone paradox. As shown in Figure 5.4, the Californians tended to have peak pitch classes in the range B, C, C♯, D, and D♯, whereas the group from the south of England tended to have peak pitch classes in the range F♯, G, and G♯.[5]

Later, in March 1997, BBC Radio conducted a large-scale experiment on the tritone paradox in a "Megalab"—a collection of experiments carried out during Science Week, with radio listeners as subjects. Listeners heard four tritones, and then phoned in to the radio station to report for each pair of tones whether it had

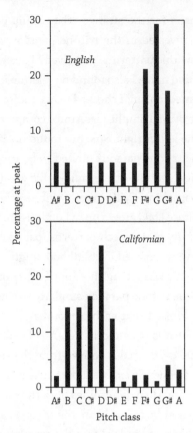

FIGURE 5.4. Perception of the tritone paradox by Californian listeners, and by listeners from the south of England. The graph shows the percentage of times that each note was a peak pitch class, for the two groups of listeners. From Deutsch (1991).

ascended or descended in pitch. Several hundred responses were collected, and they clearly supported the findings in my lab from listeners who had grown up in the south of England.

Other laboratories have since uncovered further geographic associations. Listeners in Boca Raton, Florida, produced results that were similar to those I had found among Californians.[6] In contrast, listeners in Hamilton, Ontario, in Canada, obtained findings that were very similar to those I had found among those from the south of England.[7]

So, assuming that perception of the tritone paradox is determined by an acquired speech-related template, when in life does this template develop? In one study, Frank Ragozzine and I tested a group of listeners who had all grown up in the area of Youngstown, Ohio. We found strong differences between those whose parents had also grown up in the Youngstown area and those whose parents had grown up elsewhere in the United States. Since parents have a particularly strong influence on speech development, these results indicated that a person's pitch class template was likely to have been formed in childhood.[8]

In another experiment, I studied fifteen pairs of subjects, each pair consisting of a mother and her offspring. The offspring were mostly children, but some were young adults. They were all Californian, but their mothers had grown up in many different geographic regions, including England, the European continent, and various parts of the United States. As expected from their different geographic origins, the mothers perceived the tritone paradox in ways that were strikingly different from each other. And although the offspring were all Californian, their perceptions corresponded closely to those of their mothers, and so also differed considerably from each other.[9]

We can then ask what happens in the case of young adults, who had been exposed to one language in infancy and later acquired a different language. Will they hear the tritone paradox in accordance with their first language, or will they hear it in accordance with the language that they now speak?

To explore this question, my colleagues and I tested two groups of Vietnamese immigrants. The first, younger, group had arrived in the United States as infants or young children. They all spoke perfect English, and most were not fluent speakers of Vietnamese. The second, older, group had arrived in the United States as adults. They all spoke perfect Vietnamese, but little English. We also tested a third group, consisting of English-speaking Californians, whose parents were also English-speaking Californians.

The judgments of both Vietnamese groups were clearly different from those of the native Californians, yet they didn't differ significantly from each other. In a further experiment, we asked a group of Vietnamese listeners to read out a five-minute passage in Vietnamese, and we found a strong correspondence between the pitch ranges of their speaking voices and how they heard the tritone paradox. So these results strongly support the view that perception of this illusion reflects a speech-related template that develops early in life, and survives into adulthood.[10]

A few people, on hearing the tritone paradox, have initially reported that they can hear a given pair of tones ascending and descending at the same time. When this happens, I ask them whether one direction of pitch motion appears to be the more salient, and they have always expressed a preference. Also, the impression of two simultaneous directions of motion is reduced when the listener has heard many such examples, and it can finally disappear, so that ultimately only one direction of motion for each tritone pair is heard.

◦◦

How strong is the evidence that the pitch range of a person's speaking voice is determined by the voices that he or she has been hearing in the long term? Popular opinion holds that the pitch range of speech is physiologically determined, and serves as a reflection of body size—that big people tend to have deep voices, and that small people have high ones. Yet taking male and female speech separately, the overall pitch level of a person's speaking voice doesn't correlate with the speaker's body dimensions, such as height, weight, chest size, or size of larynx (the voice box).[11] In addition, the pitch level of speech varies with the speaker's language.[12] Yet findings comparing different languages are difficult to interpret, since words and patterns of intonation vary across languages. To show that the pitch range of speech is indeed culturally determined, one needs to compare two groups of people who speak the same language; use a similar dialect; who are overall similar in appearance, ethnicity, and lifestyle; yet interact very little.

**Speech in Two Chinese Villages**

http://dianadeutsch.net/ch05ex02

I discussed this issue with my graduate student Jing Shen, and her previous advisor at East China Normal University, Jinghong Le. We concluded that we would need to find two relatively remote villages that were geographically close to each other but difficult of access—so there would be little communication between the residents of the two villages. We would then compare the pitch ranges of speech in these two villages.

The villages that we decided on—Taoyuan Village and Jiuying Village—are in a mountainous region in central China, at the border of Hubei and Chongqing. They are less than 40 miles apart, but since the roads linking them are treacherous in places, travel time between them by car takes several hours. The residents of the two villages speak in similar dialects, both being versions of Standard Mandarin, and they can understand each other reasonably well; however, given the difficulties involved in traveling between the villages, they interact very little.

We tested women who had spent most of their lives in their respective villages, by giving them the same three-minute passage of Mandarin to read out. We found a clear clustering of pitch values in the speech we recorded within each village, and a strong difference in the distributions of pitch values between the villages. This difference is illustrated in the "Speech in Two Chinese Villages" module, which contains recordings of speech from women in the two Chinese villages. So this study indicates strongly that the pitch range of a person's speaking voice is determined by long-term exposure to the speech of others.[13]

Why should dialects, including the pitch range of the speaking voice, have de-veloped in such a detailed and specific fashion? Tecumseh Fitch proposed that hu-mans initially lived in small groups or tribes that were defined by kinship.[14] From an evolutionary perspective, survival fitness is a matter not only for the individual, but also for others who share his or her genetic material. Now, the existence of dia-lects makes it easier for people who share kinship to recognize each other, and so to transmit valuable information more selectively among themselves. On this line of reasoning, incorporating pitch ranges into dialects should provide easily identi-fiable cues concerning whether a given speaker is a member of the same group or a stranger. So, viewed from an evolutionary perspective, a clustering of pitch ranges within a linguistic community would have been considerably advantageous to early humans. This advantage would have diminished with time due to increased travel between populations; however, the pitch range of a person's speaking voice still provides important clues concerning his or her place of origin.

Fitch points out that our ability to recognize dialects is far in excess of what is needed for communication. This pronounced ability would be inexplicable, he writes, if language served solely for the purpose of information exchange. He ar-gues that fitness is not just a matter of survival of the individual, nor of his or her success in producing offspring, but also of the reproductive success of all those individuals who share his or her genes. So when an individual's actions are con-ducive to the survival or reproduction of a relative (while incurring only a minor cost to himself), overall inclusive fitness is increased. From a genetic perspective, the closer the relationship between two people, the more each one has to gain by helping the other. Fitch cites the biologist J. B. S. Haldane, who recognized the evolutionary advantage of saving a drowning relative at the risk of one's own life, since he or she would be likely to share some of the relative's genes. He is said to have joked that he would "give his life for two brothers or eight cousins," recog-nizing that we share half of our genetic material with our brothers and sisters, and an eighth with our full cousins. On this line of reasoning, recognizing kin on the basis of a similar dialect—including pitch range—serves a useful purpose: the closer the kin, the more similar the dialect is likely to be.

An agreed-upon pitch range of the speaking voice is advantageous in other ways also. For example, it's very useful to determine the emotional tone of a speaker from his or her voice. When I answer the telephone, and the person at the other end of the line is someone whom I know well, I rapidly obtain a good idea of their mood from the pitch range of their voice—if it's lower than normal, they may well be feeling depressed, and if it's higher, they are likely to be excited. This would only be possible if I had a strong memory for the general pitch range of the person's speaking voice. In relation to this, the speech of adult mothers and

FIGURE 5.5. An "Inversion" produced by Scott Kim. Viewed right-side up we see letters producing the phrase "upside down." And when we turn the page upside down, we see the same thing! From Kim (1989), p.29. Copyright © 2018 Scott Kim scottkim.com.

their daughters is often strikingly similar. When my daughter was a teenager and lived at home, our voices sounded indistinguishable to people who called over the telephone.

In general, the tritone paradox provides further evidence that we do not passively analyze incoming sound patterns. Rather, we interpret these ambiguous tritones in terms of an acquired pitch class template, and our interpretation strongly influences what we perceive—in this case either an ascending pattern or a descending one. The tritone paradox therefore joins other illusions that we have been exploring, including the scale illusion and its variants, the mysterious melody illusion, and other illusions we shall be exploring later—the phantom words illusion and the speech-to-song illusion—in showing strong effects of our memories and expectations in determining how we hear musical patterns.

Put another way, the tritone paradox illustrates a perceptual principle that occurs in vision as well as hearing. When we perceive an ambiguous array, we interpret it so that it makes most sense to us. The artist and puzzle maker Scott Kim showed this by creating amazing designs that he termed *inversions*—words and phrases that are composed of letters that are slightly modified so that when you turn the page upside down, you see the same word or phrase. One of Kim's "inversions" is *upside down,* shown in Figure 5.5.[15] After viewing this figure right-side up, turn the page upside down, and behold—you perceive the same thing! We orient this array perceptually so that it is most intelligible to us, given our long experience with English letters and words. In the tritone paradox, our knowledge and experience of the pitch range of speech that we have most often heard causes us to orient this pattern in accordance with this familiar pitch range. So both Kim's "inversions" and the tritone paradox illustrate the power of top-down processing on the way we orient our perceptual world.

⟿

Finally, the tritone paradox has important implications for absolute pitch—defined as the ability to name a musical note when it's presented in isolation. This ability is very rare in the Western world. Yet when people make orderly judgments of the tritone paradox, they must be using an implicit form of absolute pitch, since they hear tones as higher or as lower depending only on their pitch classes, or note names. The tritone paradox shows, therefore, that most people possess an implicit form of absolute pitch, even when they are unable to name the notes that they are judging.[16] We now continue this theme by exploring absolute pitch and its correlates in detail.

# 6

## The Mystery of Absolute Pitch

### A RARE ABILITY THAT INVOLVES BOTH NATURE AND NURTURE

Furthermore, I saw and heard how, when he was made to listen in another room, they would give him notes, now high, now low, not only on the pianoforte but on every other imaginable instrument as well, and he came out with the letter of the name of the note in an instant. Indeed, on hearing a bell toll or a clock, even a pocket-watch, strike, he was able at the same moment to name the note of the bell or time-piece.[1]

IN THE SUMMER of 1763, the Mozart family embarked on the famous tour of Europe that established the young composer's reputation as a musical prodigy. Just before they left, an anonymous letter appeared in the *Augsburgischer Intelligenz-Zettel* describing seven-year-old Wolfgang's extraordinary abilities, not the least his absolute pitch.

People with absolute pitch (also known as "perfect pitch") name musical notes as immediately and effortlessly as most people name colors. Yet this ability is very rare in our culture—its prevalence in the general population has been estimated as less than one in ten thousand.[2] Because of its rarity, absolute pitch is often regarded as a mysterious talent that's enjoyed by only a few gifted people. This impression is supported by the fact that many famous musicians possess (or have possessed) the ability—for example Arturo Toscanini, Arthur Rubenstein, Itzhak Perlman, Yo-Yo Ma, and Joshua Bell, to name a few.

Now, in contrast to the rarity of absolute pitch, the ability to judge one musical note in relation to another is very common. So, for instance, if you play the note

FIGURE 6.1.  Painting in 1763 of Wolfgang Amadeus Mozart at age 6.

F to a musician and say to her, "This is an F," she'd have no difficulty naming the note that's two semitones higher as a G, the note four semitones tones higher as an A, and so on. (A semitone is the pitch relation formed by two adjacent notes on a keyboard, such as C followed by C♯.) What most people, including most musicians, can't do is name a note when they hear it out of context.

When you think about it, it's puzzling that absolute pitch should be so rare. Take the way we name colors for comparison. Let's suppose you showed someone a blue object and asked him to name its color. And suppose he replied, "I can recognize the color—I can distinguish it from other colors—but sorry, I just can't name it." And suppose you then showed him a red object and said, "This one is red," and he replied, "Well, since this is red, the first one must be blue", the whole interchange would be very peculiar. Yet from the viewpoint of someone with absolute pitch, this is exactly how most people name pitches—they judge the relationship between the pitch to be named and the pitch of a different note whose name they already know.

As another point, most people can easily name well-known melodies when they hear them—and naming a single note should be far easier than naming a melody.

So the real mystery concerning absolute pitch is not why some people possess it, but rather why it's so rare. It's as though most people have a syndrome with respect to labeling pitches that's like the rare syndrome of color anomia, in which patients can recognize colors, and can discriminate between different colors, but cannot associate them with verbal labels.[3] Of course, musicians are far more likely than people without musical training to have absolute pitch, yet this ability is also rare among musicians in our culture, even though they spend many thousands of hours reading musical scores, playing the notes they read, and hearing the notes they play.

The mystery becomes even deeper when we consider that most people possess an implicit form of absolute pitch, even though they're unable to label the notes they are judging. This is shown, for example, by the tritone paradox—an illusion we explored in Chapter 5. In addition, many musicians who don't have absolute pitch can nevertheless tell to a large extent whether a familiar piece of music is being played in the correct key. In one study, musically trained subjects, most of whom were not absolute pitch possessors, listened to excerpts of Bach preludes, and almost half of them could distinguish the correct version from one that had been transposed by a semitone.[4] Later studies showed that the ability to judge the pitch level of familiar music is quite widespread. Daniel Levitin asked university students to choose a CD containing a popular song they knew well (for example, "Hotel California" by the Eagles, or "Like a Prayer" by Madonna) and then to hum, whistle, or sing the tune.[5] They were tested on two different songs, and the first notes the subjects produced were compared with the equivalent notes on the CD. It turned out that 44 percent of the subjects came within two semitones of the correct pitch for both songs. In a further study by Glenn Schellenberg and Sandra Trehub, students without absolute pitch heard five-second excerpts of instrumental soundtracks from familiar TV shows. The excerpts were played either at the original pitch levels, or as transposed upward or downward by one or two semitones, and the students were able to identify the correct excerpts beyond chance.[6]

Further intriguing studies have pointed to implicit absolute pitch for individual tones. For example, the dial tone for landline telephones is ubiquitous in the United States and Canada, so listeners in this region have had thousands of experiences with this tone. When students without absolute pitch were presented with versions of the dial tone that were either correct or pitch-shifted, they identified the correct version beyond chance.[7] Another study took advantage of the fact that a 1000 Hz sine tone (a "bleep") is employed to censor taboo words in broadcast media. The "bleep" was either at the correct frequency, or it was one or two semitones removed, and (as with the dial tone study), listeners who didn't possess absolute pitch could still select the correct version above chance.[8]

Implicit absolute pitch is present very early in life, before speech is acquired. In an ingenious experiment, eight-month old infants were familiarized with streams of tones that sounded continuously. Embedded in these streams were subsequences that were repeated exactly, as well as other subsequences that had been transposed to different pitch ranges while preserving the relationships between the tones. The infants showed interest in the sequences that contained subsequences in which the actual tones were repeated, but not in those that contained transposed subsequences. Since the babies were unable to speak, they couldn't have been labeling the tones, but this study showed that they were nevertheless sensitive to their absolute pitches.[9]

Given that most people have implicit absolute pitch, its rarity as it's conventionally defined is unlikely to be due largely to limitations in pitch memory. I'll be arguing that the problem basically involves assigning verbal labels to pitches, and that we need to turn to speech and language to obtain an understanding of this baffling ability.

There are several competing views as to why some people develop absolute pitch, while most others do not do so. One view is that absolute pitch is available only to rare people who have a particular genetic endowment, and that it becomes apparent as soon as circumstances allow. Indeed, the ability often shows up at a very young age—even when the child has had little or no formal musical training. The great pianist Arthur Rubenstein describes how, when a very young child, he would listen in to his older sister's piano lessons:

> Half in fun, half in earnest, I learned to know the keys by their names, and with my back to the piano I would call the notes of any chord, even the most dissonant one. From then on, it became "mere child's play" to master the intricacies of the keyboard.[10]

Extraordinary musicians such as Mozart and Rubenstein were not the only ones to develop absolute pitch in early childhood. I vividly remember my astonishment when, at age four, I discovered that other people (even grown-ups!) couldn't name a note that I played on the piano—to do so, they needed to come over to the piano and see what key was being played. Presumably I had received some informal musical training at the time, but it would have been minimal.

Another argument for the genetic view is that absolute pitch often runs in families. In a survey of musicians, respondents with absolute pitch were four times more likely than others to report that they had a family member who possessed the ability.[11] But this argument has been criticized on several grounds. Many be-haviors that aren't genetically determined run in families—including the dialects we speak and the food we eat. As we'll be discussing shortly, the probability of acquiring absolute pitch is closely associated with beginning music lessons at an early age, and parents who provide one of their children with early music lessons are likely to do this for their other children also. Furthermore, babies who are born into families that include absolute pitch possessors are more likely to hear family members sound musical notes together with their names—so they would have the opportunity to attach names to pitches during the period in which they learn to name the values of other attributes, such as colors. At all events, there is an ongoing search for a DNA marker for absolute pitch, though so far it has proved elusive.

Other researchers have taken the opposite view—that absolute pitch can be acquired by anyone at any time, provided they practice enough. If you Google "perfect pitch training" on the Web, you'll find a large number of sites that offer programs to train you to develop the ability. But scientific attempts to train ab-solute pitch in adults have almost always come up dry. Paul Brady published one reliable report of partial success.[12] He subjected himself to a heroic course of listening to training tapes for about 60 hours, following which he obtained a score of 65 percent correct on a test of absolute pitch. Brady's report is impres-sive, but it also underscores the extreme difficulty of acquiring this ability in adulthood, in contrast to its effortless and mostly unconscious acquisition in early childhood.[13]

A third view is that acquiring absolute pitch requires exposure to certain envi-ronmental influences during a particular period early in life, known as a *critical period*. Possessing this ability is associated with early musical training—and the earlier the onset of training, the stronger the association. In large-scale direct-test studies at several music conservatories and universities in the United States and China, we found that the earlier a student had begun taking music lessons, the greater was the probability that he or she possessed absolute pitch.[14]

As was first pointed out by Eric Lennenberg in his influential book *Biological Foundations of Language*,[15] young children and adults acquire language in quali-tatively different ways. Children learn to speak automatically and effortlessly, whereas learning a second language in adulthood is self-conscious and labored—even after many years of experience, adults who acquire a second language speak it with a "foreign accent," and often with grammatical errors. So Lennenberg pro-posed that there exists a critical period for acquiring speech and language, which is at its peak from infancy to around age five, gradually declines during later child-hood, and levels off at around puberty.

Since Lennenberg advanced this conjecture, evidence for it has burgeoned. Findings from neuroscience have shown that the brain sorts out relevant from irrelevant information very early in life. Particularly during this period, connections between neurons are formed, others are strengthened, and yet others are pruned.[16] These neural changes depend on input from the environment, and they have profound and lasting effects on later function. For example, ducklings follow the first moving object they see after they hatch, and this object is most likely to be their mother. Further, white-crowned sparrows and zebra finches learn their mating songs from their fathers during a fixed period early in life. Where human speech is concerned, as Lennenberg had surmised, the period during which children most readily acquire speech and language extends up to age five or so. Proficiency in acquiring speech gradually decreases after this age, and following puberty, the ability to acquire language for the first time becomes seriously impaired.

One type of evidence that acquiring language involves a critical period concerns children who were socially isolated in childhood, and later placed in a normal environment. After they were rescued, these children couldn't acquire normal speech despite intensive training. One child, called Genie, was locked in a small room by her deranged parents, and completely deprived of language during her entire childhood. She was discovered at age twelve by a Los Angeles social worker, and placed in a regular environment, where a team of psychologists at UCLA worked hard to rehabilitate her. Yet despite expert and extensive training, Genie never learned to speak normally.[17] Another such child was Victor of Aveyron, who was discovered in the year 1800 at about twelve years of age, having apparently spent most of his childhood in the woods. After he was found, a medical student, Jean Marc Gaspard Itard, adopted him into his home and made extensive attempts to teach him to talk. Victor learned a few words, but failed to acquire normal speech, and after several years Itard gave up his attempts.[18,19]

Studies of speech recovery following brain injury that occurred at different ages also point to a critical period: Recovery is best when the injury occurs before age six, declines after that, and is very poor following puberty.[20] Further evidence concerns deafness. A very small percentage of deaf children are born to deaf parents who use sign language, and these children learn to sign in the same way as hearing children learn to speak. But most deaf children are born to hearing parents who do not sign, and they generally begin to acquire sign language when they attend schools for the deaf. Elissa Newport compared signers who were exposed to sign language from birth with those who were first exposed to it at ages 4 - 6, and also with those who were first exposed to it after age twelve. She found that the earlier her subjects had been exposed to sign language, the better they performed on certain linguistic tests.[21]

Other related research involves second-language learning. Since much of the basic brain circuitry for speech is established during acquisition of a first language, acquiring a second language doesn't depend vitally on a critical period. But even then, there's a strong relationship between the age of learning a second language and how successful it is. Studies of immigrant populations have shown that when this learning occurs early in life (up to 4–6 years), the child learns to speak the language as well as a native. But later, the ability to learn the language gradually declines, and it reaches a plateau in adulthood.[22] This is particularly true of the sound patterns of speech, so that even following many years of experience, adults who acquire a second language generally speak it with a "foreign accent." The author Joseph Conrad was a striking case in point. Born in the Ukraine, he spoke Polish as his first language. At age sixteen he moved to Marseilles, where he joined the merchant marine and soon learned to speak French with a southern French accent. At age twenty he moved again, this time to England, where he rapidly became so proficient in the English language that he became one of the greatest writers of English literature. But he spoke English with such a strong French accent that people had considerable difficulty understanding him, and he avoided speaking in public. (His son, Borys Conrad, was once startled to hear him pronounce *iodine* as "you're a-dyin").[23]

Tone languages, such as Mandarin, Cantonese, and Vietnamese, strengthen the case for a link between absolute pitch and speech. In these languages, words take on entirely different meanings depending on the *lexical tones* in which they are spoken. Lexical tone is defined both by pitch height and pitch contour. In Mandarin, for example, the first tone is high and level, the second is mid-high and rising, the third is low, initially falling and then rising, and the fourth is high and falling. This is illustrated in the accompanying module. The word *ma* when spoken

**A Mandarin Word in Four Different Tones**

http://dianadeutsch.net/ch06ex01

in the first tone means *mother*; in the second tone it means *hemp*; in the third tone it means *horse*; and in the fourth tone it means a reproach. So pitches in tone languages are used to create verbal features, analogous to consonants and vowels. Therefore when speakers of Mandarin hear the word *ma* spoken in the first tone and attribute the meaning *mother*, or when they hear *ma* spoken in the third tone and attribute the meaning *horse*, they are associating a pitch—or a sequence of pitches—with a verbal label. Analogously, when people with absolute pitch identify the sound of the note F♯ as *F♯*, or the note B as *B*, they are also associating a pitch with a verbal label.

So when babies hear Mandarin being spoken, they automatically associate pitches with words, so they develop absolute pitch for the words they hear. Therefore when they reach the age at which they can begin taking music lessons, their brain circuitry for absolute pitch is already in place, so they develop absolute pitch for musical tones similarly to the way they would acquire words in a different tone language. In contrast, nontone languages such as English use pitch for purposes such as conveying grammatical structure, emotional tone, and other aspects of prosody, but not for determining the meaning of individual words. In consequence, speakers of nontone languages are at a disadvantage for acquiring absolute pitch.

I first realized that there might be a relationship between absolute pitch and tone language when I was testing a group of native Vietnamese speakers in an unrelated experiment. Many of my subjects were recent immigrants, and they spoke very little English. I was intrigued by the musical quality of their speech, and tried to imitate it. But when I first attempted to repeat a Vietnamese word back to them, they thought that I was either talking nonsense or that I was saying a word with an entirely different meaning. So I tried varying the pitch in which I pronounced a word—going up and down a scale—and was astonished to find that, depending on how high or low I pitched my voice, the subjects would give the word an entirely different meaning.

Because pitch turned out to be important for conveying meaning in Vietnamese, it seemed to me that native speakers of this tone language must be using absolute pitch in the process of understanding the meaning of words. If this were so, then it would follow that they'd be very consistent in the pitches with which they pronounced a list of words on different days. To find out, I tested seven native Vietnamese speakers,

**Pitch Consistency in Tone Language Speech**

http://dianadeutsch.net/ch06ex02

none of whom had received any significant musical training. I handed them a list of ten Vietnamese words to read out on two different days. Then for each speaker, my colleagues and I calculated the difference between the average pitch of each word as it was read out on the different days, and we averaged these differences across the words in the list.[24] Our results showed remarkable consistencies: Four of the seven speakers showed averaged pitch differences across days of less than half a semitone, and for two of these speakers, the averaged pitch discrepancies were less than a quarter of a semitone—an amazingly small difference!

We next wondered whether speakers of a different tone language would show the same pitch consistencies in pronouncing lists of words. To find out, I tested fifteen speakers of Mandarin, who had also received little or no musical training. I generated a list of twelve Mandarin words, with each of the four tones occurring three times in the list.

I tested the Mandarin speakers in two sessions, which were held on different days, and they read out the word list twice in each session. On analyzing the pitches of the words, we again found remarkable consistencies: Half of the subjects showed averaged pitch differences of less than half a semitone, and one-third of the subjects showed averaged pitch differences of less than a quarter of a semitone! And importantly, there were no significant differences in the degree of pitch consistency with which they recited the word list on the different days, compared to reciting it twice in immediate succession. The module "Pitch Consistency in Tone Language Speech" presents, as examples, the word lists recited by Vietnamese and Mandarin speakers on different days.

How do speakers of English perform on this task? To find out, I tested fourteen English speakers on a list of English words, using the same procedure. The English speakers showed roughly the same degree of pitch consistency as the Mandarin speakers when they recited the word list twice in immediate succession, but they were significantly less consistent in reciting the same list on different days.

These findings led me to conclude that speakers of Vietnamese and Mandarin possess a remarkably precise form of absolute pitch for the tones of their language, which was here reflected in the way they pronounce words. So I conjectured that this ability resulted from their acquisition of tone language in infancy, when they had learned to associate pitches with meaningful words. I presented these findings at a 1999 meeting of the Acoustical Society of America, in a paper boldly titled "Tone Language Speakers Possess Absolute Pitch," and this was featured on the next day on the front page of the *New York Times*.[25] In this paper I argued that absolute pitch, which had traditionally been viewed as a musical ability, originally evolved to subserve speech, and that it is still reflected in the speech of East Asian tone languages. I also conjectured that when infants acquire absolute pitch as a feature of the tones of their language, and they later reach the age at which they begin taking music lessons, they acquire absolute pitch for musical tones in the same way as they would acquire the tones of a second tone language. In contrast, children who first acquire a nontone language such as English would need to learn to name the pitches of musical tones as though they were the tones of a first tone language—so they would perform very poorly on tests of absolute pitch. Also, based on the findings showing a critical period for acquiring language, I conjectured that both tone language and nontone language speakers would show a clear effect of the age at which they had begun taking music lessons.

The obvious next step, then, was to compare two groups of people on a test of absolute pitch for musical tones. One group would be speakers of English, and the other group would be speakers of a tone language, such as Mandarin. But when I first attempted to set up this study, I encountered a surprising obstacle: The English-speaking researchers with whom I discussed the idea said it would be a waste of time, since absolute pitch was too rare for us to obtain a significant result.

**Brief Test for Absolute Pitch**

http://dianadeutsch.net/ch06ex03

And the Chinese researchers whom I approached also said that the study would be pointless, but for a different reason: Many fine musicians in China have absolute pitch anyway. But I finally found two enthusiastic collaborators who agreed to test students at renowned music conservatories in the US and in China: Elizabeth Marvin, then dean of the Eastman School of Music in Rochester, New York, tested students at Eastman, and HongShuai Xu, then at Capital Normal University in Beijing, tested students at the Central Conservatory of Music there.

At Eastman, we gave a test of absolute pitch to 115 first-year students—all English speakers. And at the Central Conservatory in Beijing, we gave the identical test to 88 first-year students—all Mandarin speakers. We used a score of at least 85 percent correct as our criterion for possessing absolute pitch. (You can test yourself by naming the notes presented in the "Brief Test for Absolute Pitch" module.

The students' judgments provided a remarkable confirmation of my speech-related critical period hypothesis: For both the Mandarin and the English speakers, the earlier they had begun taking music lessons, the higher their probability of meeting the criterion for absolute pitch. And for all levels of age of onset of musical training, the percentage of those with absolute pitch was far higher for the Mandarin than for the English speakers. For students who had begun taking music lessons at ages 4 and 5, approximately 60 percent of the Mandarin speakers met the criterion, while only 14 percent of the English speakers did so. For students who had begun music lessons at ages 6 and 7, roughly 55 percent of the Mandarin speakers met the criterion, whereas this was true of only 6 percent of the English speakers. For those who had begun music lessons at ages 8 and 9, approximately 42 percent of the Mandarin speakers met the criterion, whereas none of the English speakers did so. And as shown in Figure 6.2, when we relaxed the

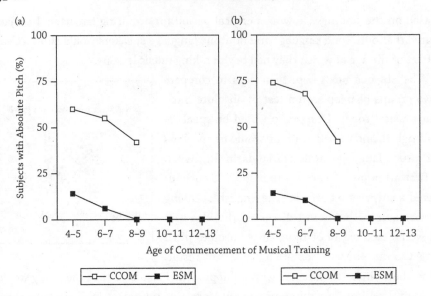

FIGURE 6.2. Percentages of subjects who obtained a score of at least 85 percent correct on our test for absolute pitch, as a function of age of onset of musical training. (A) No semitone errors allowed. (B) Semitone errors allowed. Unfilled boxes show the results from students at the Central Conservatory of Music (CCOM) in Beijing, China; these students were all tone language speakers. Filled boxes show the results from students at the Eastman School of Music (ESM), in Rochester, New York; none of these were tone language speakers. From Deutsch et al. (2006).

criterion for absolute pitch by allowing for semitone errors, the difference between the Central Conservatory and the Eastman students was even more pronounced.[26]

These findings strongly support my conjecture that when infants acquire a tone language, they develop absolute pitch for the tones of this language, so they can later acquire absolute pitch for musical tones in the same way as they would acquire the tones of a different tone language. This would explain why most speakers of English, who don't have the opportunity to associate pitches with meaningful words in infancy, find it far more difficult to acquire absolute pitch for musical tones, even in childhood.

The evidence that I've described for this view applies so far only to speakers of tone languages, in which pitch is prominently involved in determining the meaning of words. But the same principle also applies to some extent to other East Asian languages such as Japanese and certain dialects of Korean. Japanese is a pitch-accent language, in which the meaning of a word changes depending on the pitches of its constituent syllables. For example, in Tokyo Japanese the word *hashi* means *chopsticks* when it is pronounced high-low; *bridge* when it is pronounced low-high; and *edge* when there is no pitch difference between its two syllables. In Korean, the Hamkyeng and Gyeonsang dialects are considered tonal or pitch accent. For example, in the South Gyeonsang dialect, the word *son* means *grandchild* or *loss* when it is spoken in a low tone, *hand* in a mid tone, and *guest* in a high tone. This leads to the conjecture that people who had been exposed to these languages or dialects in infancy might also have a higher prevalence of absolute pitch, yet this prevalence would not be as high as among people who speak a tone language. And indeed, it has been shown that absolute pitch is more prevalent among speakers of Japanese than among nontone language speakers, but less prevalent than among speakers of Mandarin.[27]

∾

As an alternative explanation of the results of our Eastman/Beijing study, the differences we found between the Chinese and American students could have been based on genetics or ethnicity rather than language. To examine this possibility, my colleagues and I carried out a further study.[28] We gave the same test for absolute pitch to first- and second-year students who were taking a required course at the University of Southern California (USC) Thornton School of Music. Based on their responses to a questionnaire, we divided the students into four groups: Those in group *nontone* were Caucasian, and spoke only a nontone language such as English fluently. The remaining students were all of East Asian ethnic heritage (Chinese or Vietnamese). Those in group *tone very fluent* spoke a tone language "very fluently." Those in group *tone fairly fluent* spoke a tone language "fairly fluently." And those in group *tone nonfluent* responded, "I can understand the language, but don't speak it fluently." Then we divided each of the language groups into subgroups by age of onset of musical training. One subgroup had begun training at ages 2–5 years, and the other had begun training at ages 6–9 years.

As shown in Figure 6.3, all language groups showed clear effects of age of onset of musical training—in all cases, the average scores for those who had begun training between ages 2–5 were higher than for those who had begun training between ages 6–9. But there was also an overwhelming effect of tone language fluency, with ethnicity held constant: Those who spoke a tone language *very fluently* produced remarkably high scores, mostly in the 90 percent to 100 percent correct region. Their scores were far higher than those of the Caucasian *nontone* speakers, and importantly, they were also far higher than those of East Asian ethnicity who were *tone nonfluent*. The scores of the *tone very fluent* speakers were also higher than those of the *tone fairly fluent* speakers, whose scores were in turn higher than those of the *tone nonfluent speakers,* as well as of the *nontone* speakers. So our results showed that the differences in performance between the tone and nontone language speakers on our absolute pitch test were associated with language rather than with ethnicity.

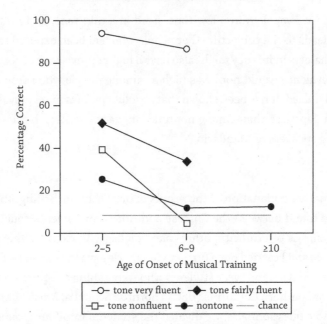

FIGURE 6.3. Average percentage correct on a test of absolute pitch among students in a large scale study at an American music conservatory. The data are plotted as a function of age of onset of musical training, and fluency in speaking a tone language. Those in groups *tone very fluent*, *tone fairly fluent*, and *tone nonfluent* were all of East Asian ethnic heritage, and spoke a tone language with differing degrees of fluency. Those in the group *nontone* were Caucasian and spoke only nontone language. The line labeled "chance" represents chance performance on the task. From Deutsch et al. (2009).

Another suggested explanation for the disparities we obtained between these groups is that they resulted from differences in music teaching practices in the US and China. We examined this possibility by comparing the students at USC Thornton who had arrived in the US from China after age nine—so would have received their early music education in China—with those who had been born in the US, or who had arrived in the US before age nine—so their music education would have been in an American system. We also compared the *tone very fluent* students at USC Thornton with the students at the Central Conservatory of Music in Beijing. It turned out that there were no significant differences between these groups, though surprisingly those at USC Thornton performed at a slightly higher level than those at Beijing—perhaps because they had been more consistently exposed to Western musical scales. At all events, this analysis showed that the differences we found between the tone and nontone language speakers were not due to musical training practices in the two countries. Taken together, the results of our study point overwhelmingly to language as responsible for the differences in prevalence of absolute pitch that we obtained.

An intriguing pharmacological study lent further support to the hypothesis of a critical period for acquiring absolute pitch. Studies on animals had found that valproate—valproic acid, a drug used to treat epilepsy—appeared to restore brain plasticity and so to reopen the critical period later in life. An international team of researchers led by Takao Hensch at Harvard University recruited young men with little or no musical training, and trained them to associate pitches with names.[29] One group of subjects was given valproate, and these subjects learned to identify pitches by name better than a control group who were instead given a placebo. The number of subjects in this study was small, but it will be exciting if the results hold up when larger groups are tested.

ᘓ

A number of neuroanatomical studies have also pointed to a strong relationship between absolute pitch and speech. These studies have shown that absolute pitch possessors have a unique brain circuitry, particularly involving regions in the temporal lobe that are concerned with speech processing. Most of the interest here has focused on the planum temporale—a region in the temporal lobe that is located behind the auditory cortex. In the vast majority of right-handers, the planum temporale is larger in the left hemisphere, where it forms the core of Wernicke's area—a region that is important for the comprehension of speech. In a classic MRI study, Gottfried Schlaug and his colleagues at Harvard showed that this left-wise asymmetry of the planum temporale was more pronounced among musicians who possessed absolute pitch than among those who didn't possess this ability.[30] Further neuroanatomical evidence involves white matter tracts that connect together neurons in different regions of the brain. Psyche Loui, Schlaug, and their colleagues at Harvard showed that the volumes of white matter tracts linking the left superior temporal gyrus with the left medial temporal gyrus—which are important to speech processing—were larger among absolute pitch possessors than among nonpossessors.[31]

These studies, which indicate that people with absolute pitch may have an unusually effective brain circuitry for speech, provide a clue as to why some nontone-language speakers acquire absolute pitch. Perhaps such people have unusually strong memories for spoken words, and this facilitates the development of associations between pitches and their spoken labels early in life. To test this conjecture, my colleague Kevin Dooley and I used the digit span, which measures how long a string of digits a person can hold in memory and recall in correct order. We recruited two groups of English-speaking subjects—one consisting of absolute pitch possessors, and the other consisting of nonpossessors who were matched with the first group for age, and for age-of-onset and duration of musical training. On each

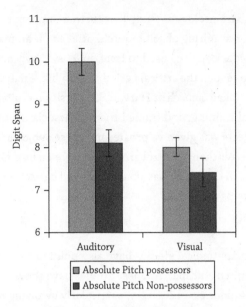

FIGURE 6.4. Average scores on the auditory and visual digit span tests produced by absolute pitch possessors and nonpossessors. The subjects were speakers of English. From Deutsch & Dooley (2013).

trial, the subject was presented with a string of digits, and attempted to repeat them back in the order in which they had occurred. The first two trials consisted of six digits, the next two trials of seven digits, the next two of eight digits, and so on. The subject's score was set as the largest number of digits that he or she reported back correctly at least once. In the first part of the experiment the digits were presented as spoken words, and in the second part they were presented visually on a computer screen.[32]

As shown in Figure 6.4, when the digits were spoken words, the absolute pitch possessors performed better than the others. In contrast, when the digits were presented visually, the two groups showed very similar performance. So our findings indicated that absolute pitch is indeed associated with an unusually large memory span for speech sounds. As the most probable explanation, a substantial memory for speech sounds should increase the likelihood that a child would develop associations between pitches and their spoken labels early in life, so promoting the development of absolute pitch. Interestingly, other studies have shown that the size of the digit span is genetically determined, at least to some extent. So, at least for speakers of nontone languages such as English, the genetic basis associated with an enhanced memory for spoken words could also contribute to the development of absolute pitch.

᷎ᴐ

Even though absolute pitch possessors can name notes with very few errors, their naming accuracy still depends on the characteristics of the notes they hear. In general, pitches that correspond to white keys on the piano—C, D, E, F, G, A, B— are identified more accurately than those that correspond to black keys—C♯/D♭, D♯/E♭, F♯/G♭, G♯/A♭, A♯/B♭. A suggested explanation for this *black/white key effect* is that most people with absolute pitch would have begun taking music lessons on the piano using only white keys, with the black keys being introduced gradually as learning progressed. In particular, Ken'ichi Miyazaki at Niigata University in Japan has argued that this white-key advantage results from practice on the piano in early childhood.[33] If this hypothesis were correct, then we would expect the advantage to be more pronounced in pianists than in other instrumentalists.

Together with my colleagues Xiaonuo Li and Jing Shen, I investigated this question in a large-scale study of students at the Shanghai Conservatory of Music.[34] We compared two groups: Those in the first group were pianists whose primary instrument had always been the piano, and those in the second group were orchestral performers whose primary instrument had always been orchestral. Both groups showed a black/white key effect, but this was slightly more pronounced among the orchestral performers than among the pianists. So our findings showed that the white key advantage couldn't be due to early training on the piano.

Annie Takeuchi and Stewart Hulse at Johns Hopkins University proposed a different explanation for this effect.[35] They suggested that performance on absolute pitch tests might be more accurate for pitches that the subjects had heard most frequently in the long run, and that in Western tonal music white-key pitches occur more frequently than black-key pitches. This could explain the black/white key effect, and further suggests that performance might depend in a more fine-grained fashion on the frequency of occurrence of pitches in the tonal music repertoire.

With this conjecture in mind, we continued to analyze the judgments of the students at the Shanghai Conservatory of Music, and we plotted the overall accuracy of note naming as a function of the pitch class of each note (C, C♯, D, and so on). Then we correlated these patterns of accuracy with the frequency of occurrence of each pitch class in Barlow and Morgenstern's *Dictionary of Musical Themes*. (This dictionary comprises 9,825 musical themes from the works of over 150 composers of classical music.[36]) As shown in Figure 6.5, we found that overall, the more frequently a pitch class occurred in the repertoire, the more often it was named correctly. So, for example, the note D occurred most often and was most often named correctly, and the note G♯ occurred least often and was least often named correctly. So it appears that absolute pitch is acquired and maintained for the different pitch classes independently, to some extent.

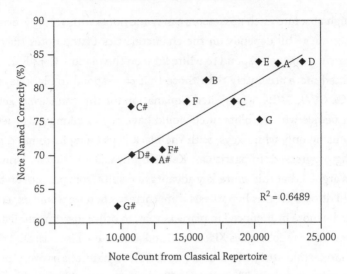

FIGURE 6.5. Average percentage correct on a test of absolute pitch among students at the Shanghai Conservatory of Music, plotted for each pitch class separately, and against the number of occurrences of each pitch class, derived from Barlow and Morgenstern's *Electronic Dictionary of Musical Themes* (2008). From Deutsch et al. (2013).

Accuracy in absolute pitch judgment also varies depending on the octave in which the note is played. In general, notes in the middle of the musical range (roughly in the two-octave range around Concert A) tend to be recognized most accurately, and accuracy drops off as the pitches diverge from the middle of this range. Perhaps this reflects differences in the frequency of occurrence of notes in different pitch ranges, but so far this suggestion hasn't been formally tested.

Does the timbre (that is, the sound quality) of a tone affect how well its pitch is identified? Some people can name pitches accurately regardless of how they are produced—they'll recognize, for example, that a car horn is sounding F♯, or that their vacuum cleaner is producing B♭. But other people, while being able to identify accurately the tones produced by familiar musical instruments, perform much less well in identifying the pitches of tones with different timbres. In one study, absolute pitch possessors were asked to identify the names of tones consisting of synthesized piano, viola, or sine waves. Piano tones were associated with the highest scores, followed by viola tones, and sine waves were identified least well.[37] In another study, tones from a number of instruments were recorded, and in some cases their initial portions were spliced out, causing their timbres to sound unfamiliar. The pitches of the truncated tones were named much less well, indicating that the sound of the tone as a whole is important to absolute pitch judgment.[38] So although the circuitry underlying absolute pitch is established early in life, the ability to name the pitch of a note is also affected by its timbre.

✌

Does the context in which a note is played affect how absolute pitch possessors perceive it? Steven Hedger and his colleagues at the University of Chicago had people with absolute pitch listen to Brahms's Symphony No. 1 in D minor. During the first 15 minutes, the researchers gradually tweaked the tuning of the recording so that it ended up 1/3 semitone flat overall. Then they played the remainder of the symphony with the notes consistently detuned. Both before and after hearing the symphony, the listeners were played violin tones and asked to name them, and also to rate whether the tones were in tune, sharp, or flat. Before hearing the detuned symphony, the listeners' ratings reflected the actual tunings of the notes. But after listening to the detuned music, their judgments changed—they rated the flat notes as in tune more often, and the in-tune notes as sharp more often. So although they continued to name the notes correctly, they altered their tuning judgments in accordance with the notes they had just heard.[39]

The experiment by Hedger and colleagues was the first to show that absolute pitch possessors can be induced to make intonation errors by the context in which notes are played. However, their subjects still named the notes correctly, so their absolute pitch, defined as the ability to name a note on hearing it, was not impaired. In a recent study, we showed that absolute pitch possessors can also be induced by context to make specific errors in naming notes. Subjects with absolute pitch listened to two test tones that were separated by a sequence of six intervening tones. The test tones were either identical in pitch or they differed by a semitone. The subjects were asked to listen to the first test tone, to ignore the six intervening tones, and then to listen to the second test tone. Their task was to write down the names of the first and second test tones after hearing the full sequence. In one condition in which the test tones differed in pitch, a tone that was identical to the second test tone was included among the intervening tones. For example, if the first test tone was D and the second test tone was D♯, the note D♯ might be included in the intervening sequence. In this condition, the absolute pitch possessors showed an increased tendency to misname the first test tone as having the same pitch as the second test tone. (In this example they tended to misname the first test tone as D♯ rather than D.) This showed that absolute pitch possessors can be induced to make errors in note naming by the context in which tones are played.[40]

How does absolute pitch relate to overall musical proficiency? This has been the subject of many heated arguments, with some people insisting that it reflects exceptional musical ability, and others taking the opposite view—a few even claiming that it provides a musical disadvantage. The second view can surely be discounted, given that many world-class musicians possess the ability. And absolute pitch is evidently not a requirement for exceptional musicianship, since other world-class musicians, such as Stravinsky, are believed not to have possessed it. Exceptional cases aside, it's clear that the term "musical ability" is used very loosely, and in reality refers to a complex of abilities that don't necessarily correlate with each other. So some studies have explored relationships between absolute pitch and the ability to perform other musical tasks.

In one study, Kevin Dooley and I gave subjects a musical dictation test that was modeled after the placement examination administered to entering music majors at the USC Thornton School of Music. We recruited two groups of listeners—possessors and nonpossessors of absolute pitch, who were matched for age, and for age of onset and duration of musical training. We played the listeners three short musical passages to notate, furnishing them with the starting note of each passage. Then for each listener we calculated the percentage of tones that were correctly notated in their correct order in all the passages. The results showed a clear advantage to the absolute pitch possessors—the better a listener scored on our absolute pitch test, the better he or she scored on musical dictation.[41] In a further experiment we presented two such groups of listeners with melodic intervals that ranged from one to twelve semitones, and asked them to name the intervals. The absolute pitch possessors performed very well on this task, considerably outperforming the nonpossessors.[42] So these experiments showed that absolute pitch is correlated with other musical advantages.

ᕁ

Absolute pitch possessors who are getting on in years are often puzzled by a gradual deterioration in their ability to name notes in isolation—beginning at age 40 to 50, they commonly find that pitches appear to be slightly sharper or flatter than before. These pitch shifts may be related to changes in the tissues of the inner ear, which inevitably alter with age.

The British psychologist Philip Vernon described his own experiences with this pitch shift in detail. At age 52 he noticed a tendency to identify keys a semitone higher than they should be. As he wrote:

This was highly disconcerting, since I also tend to associate particular keys with particular moods—a kind of synaesthesia . . . C major to me is strong and masculine, whereas C sharp is more lascivious and effeminate. Hence it was distressing to hear, for example, the Overture to Wagner's *Mastersingers of Nuremberg*—an eminently C major composition—apparently played in C sharp. Thus when I was singing in church, or a chorus, or listening to music at concerts, I habitually transposed a good deal of the time. That is, I inferred that what I heard as C sharp or D was really C and C sharp respectively.

Now at the age of 71 the inevitable seems to be happening, and I usually identify notes or keys 2 semitones (one whole tone) too high. Wagner's Overture is now quite clearly being played in the key of D![43]

For Sviatoslav Richter, one of the great pianists of his day, absolute pitch was integral to music as he heard and played it. He was devastated in his later years to find that his absolute pitch had altered. He wrote:

I used to have perfect pitch and could reproduce everything by ear, but I noticed that my hearing is getting worse. Today I mix up keys and hear things a tone or sometimes two whole tones higher than they actually are, except for the bass notes, which I hear as being lower than they are, as the result of a kind of softening of the brain and weakening of the auditory system, as though my hearing were out of tune. Before me, Neuhaus and Prokofiev were both afflicted by a similar phenomenon—in Prokofiev's case, he heard everything up to three whole tones too high towards the end of his life. It's sheer torture. . . .[44]

Shifts in absolute pitch judgment can also result from various medications. For example, carbamazepine—a drug known as Tegretol, that is widely used to treat epilepsy and other disorders—produces a temporary downward pitch shift of roughly a semitone. In a detailed study, a concert pianist was given either carbamazepine or a placebo. She listened to piano tones, and for each tone she adjusted a bar on a computer screen to indicate the pitch she perceived. The results were remarkably orderly, with carbamazepine producing a downward shift that was on average a little below a semitone, and with the extent of the pitch shift increasing systematically from the lowest to the highest octave over a six-octave range.[45]

In Western countries, although the prevalence of absolute pitch is generally very low, it is considerably higher in certain populations. Specifically, absolute pitch is particularly prevalent in blind musicians—whether their blindness is congenital or caused by some event that occurred early in life. Famous blind musicians with absolute pitch include Stevie Wonder, Ray Charles, and jazz pianist Art Tatum. In one study, 57 percent of early blind musicians were found to possess absolute pitch, and some of them had even begun taking music lessons at an age beyond the critical period.[46] Blind people often have a high level of musical proficiency in general, and they are also superior to sighted people in performing other auditory functions—for example, they are better at localizing sounds, judging the direction of pitch change, and discriminating speech in noise. There is evidence that in blind people, brain regions that are commonly associated with visual processing are recruited for tasks involving hearing.[47] In people who have become blind—even after childhood—the occipital cortex (which includes the visual cortex) is activated in response to sound, and it's believed that this neural reorganization is responsible for the superior performance of blind people in judging sounds. Interestingly, although sighted people with absolute pitch have an exaggeration of the normal leftward asymmetry of the planum temporale, this is not necessarily true of blind people with absolute pitch.[46]

Absolute pitch also appears to occur with somewhat greater frequency among people with autism—a condition that is characterized by deficits in intellectual reasoning and communication. Such people often have islands of enhanced specific abilities, and the prodigious talents of these exceptional people (called *savants*) are often musical—in addition to having absolute pitch, they may be gifted composers, performers, and improvisers. Musical savants often focus on pitch in evaluating their environment. For example, one autistic person became distressed when he was traveling on the Paris Metro, because the pitch of the door-opening signal differed from the one used in the London Underground.[48]

In contrast to normal absolute pitch possessors who have an exaggeration of the left-wise asymmetry of the planum temporale, in autistic people the volume of the planum temporale is significantly reduced in the left hemisphere, so that for them this exaggerated left-wise asymmetry is lacking.[49]

ᴖ

Finally, we recapitulate various ways in which absolute pitch reflects a linkage between music and language. There is a strong relationship between the possession of absolute pitch and the language spoken by the listener—the prevalence of this ability is far higher among tone language speakers than among speakers of nontone language such as English. Related to this, there is a clear critical period for acquiring absolute pitch, and this coincides with the critical period for acquiring language. Further, absolute pitch possessors perform better than nonpossessors on tasks involving short term memory for words. In addition, several brain regions that are important to language are also involved in absolute pitch. So while other aspects of this ability have also been reviewed in this chapter, its relationship to speech is particularly significant.

In the next chapter we return to exploring the influence of our knowledge and expectations on illusions of sound perception, in this case focusing on the sound patterns of speech.

# 7

## Phantom Words

HOW OUR KNOWLEDGE, BELIEFS AND EXPECTATIONS CREATE

ILLUSIONS OF SPEECH

IN OCTOBER 2008, the Bolivar, Missouri *News* ran the following story:

> When Gary Rofkahr from Ossowo heard there was a baby doll on local store
> shelves that said "Islam is the light" he didn't believe it.
> He bought the Little Mommy Real Loving Baby Cuddle and Coo doll to see
> for himself, and was shocked to hear "Islam is the light" among other baby
> gibberish including "Mama."
> "I have a 1-year old granddaughter" he said. "It makes me mad that
> someone is trying to indoctrinate our children with an innocent toy."[1]

The Little Mommy doll is only one of a long string of talking toys that have
shocked customers over the years. Another such doll is Tinky Winky, whose utterances so shocked customers that they threatened lawsuit. In April 2008, Oceanside
APB news published the following:

> Taylor, an unemployed office worker, told APBnews.com today that she
> bought a talking Tinky Winky doll a few weeks ago for her daughter's second
> birthday. . . .
> When Taylor tugged at Tinky Winky's left hand to make him talk, the doll
> spouted words that shocked Taylor.

"He was saying, 'I got a gun, I got a gun, run away, run away,'" recalled Taylor, 22. "My gut just kind of dropped."[2]

In reality, according to the manufacturer Mattel/Fisher Price, the Baby Cuddle & Coo Doll features cooing, giggling, and baby babbling sounds, with only one scripted word, "Mama." Still, these sounds so alarmed parents that they formed a protest group, entitled MAMA (Moms Asking Mattel for Accountability), demanding that the doll be removed from stores. The second talking doll, Tinky Winky, was derived from the Teletubbies TV show. According to its manufacturer, the Itsy Bitsy Entertainment Co., this doll was in fact saying "Again, again" from a snippet of audio taken from the show.

These incidents reflect a profound truth: When listening to speech, the words and phrases that we hear are strongly influenced not only by the sounds that reach us, but also by our knowledge, beliefs, and expectations. This might seem surprising, as the process of speech communication generally appears effortless. So from our subjective experience, we can reasonably assume that the speech signal travels in a straightforward fashion from our ears to our brain, where, in a few simple steps, it is chopped up into small units that are then reassembled so that the original message is correctly perceived. Yet for half a century engineers have struggled to develop computer software that can understand conversational speech as well as humans can, and although they have made considerable progress, they have yet to achieve this goal.

As part of the puzzle, we can recognize words and phrases that are produced by different speakers—including those using different dialects—and we can also recognize words that are produced by the same speakers when they are in different emotional states. It's unclear how the brain is able to recognize that such varied speech sounds are saying the same things. In addition, speech often occurs in the presence of other sounds, and so needs to be separated out from them perceptually. As yet another complication, the spectrum[3] of a sound recording produced by a speech signal can be quite different depending on where in the room the recording is made—and even depending on the position of the speaker inside the room. It's unclear how our brains decide that the same sound is being presented when we are listening to such different versions.

So it's still a mystery how we are able to understand normal conversational speech effectively. But where we have an advantage over computers (and probably always will have) is that we can draw on experiences that occurred throughout our lives, and on our expectations that resulted from these experiences, in making inspired guesses concerning what is being said. However, this very process of guesswork can lead us to "perceive" phantom words and phrases that are not, in reality, being spoken.

There are many examples of such mishearings. For example, there are *Mondegreens*, which were popularized by Jon Carroll in his *San Francisco Chronicle* column. Here a phrase is given an entirely different interpretation, yet the sounds that are misheard are very similar to the original ones. The term "Mondegreen" was coined by the writer Sylvia Wright, who as a child heard the Scottish ballad "The Bonny Earl of Murray" and thought that one stanza went as follows:

> Ye Highlands and ye Lowlands
> Oh where hae you been?
> They hae slay the Earl of Murray,
> And Lady Mondegreen.

The last line of the stanza is in fact, "And laid him on the green."

Carroll collected Mondegreens from his readers for many years, and it appears that the most frequently submitted one was "Gladly, the cross-eyed bear" (from the hymn "Gladly the Cross I'd Bear"). Other favorites include "There's a bathroom on the right"—a mishearing of "There's a bad moon on the rise" from the song "Bad Moon Rising"; "Excuse me while I kiss this guy," in fact "'Scuse me while I kiss the sky," from the song "Purple Haze" sung by Jimi Hendrix; and "Surely Good Mrs. Murphy shall follow me all the days of my life" from Psalm 23: "Surely goodness and mercy shall follow me all the days of my life." In Bob Dylan's "Blowin' in the Wind," "The answer my friends, is blowin' in the wind" has been heard as "The ants are my friends, they're blowing in the wind." And in the fadeout of "Strawberry Fields" John Lennon says, "I am very bored," which has been heard as "I bury Paul"[4]

During the Second World War, the British government appointed the art historian Ernst Gombrich to supervise a "Monitoring Service," which kept watch on radio transmissions from both friends and enemies. Given the technology of the day, some of the transmissions were hardly audible, so the task of inferring the intercepted messages was a difficult one. In an internal memo entitled *Some Axioms, Musings, & Hints on Hearing*, Gombrich argued for a strong influence of the listener's knowledge and expectations on the interpretation of these intercepted sounds. He wrote: "It is the story of the signaler who misheard the urgent message 'Send reinforcements, am going to advance" as "Send three and four pence, am going to a dance.'"[5]

We use unconscious inference very frequently when we are watching movies. Film sound designers typically produce their effects by creating soundtracks and synchronizing them with the recorded video. Some of the effects they use will surprise you. For example, twisting and snapping celery sticks can produce a convincing effect of breaking bones. Slowly compressing packets of potato chips can create a persuasive impression of the crackling of fire.

The master sound designer Ben Burtt, who created the audio for the movie *Raiders of the Lost Ark,* described how he created the impression of slithering snakes in the Well of Souls:

There were a lot of elements that went into the Well of the Souls, and most principally were the snakes. And we started out recording some real snakes, but snakes don't really vocalize all that much. And part of the element of the snakes is really a movement over each other. And we've had a lot of luck over the years with cheese casserole. My wife makes a cheese casserole and it's in the dish and you just run your fingers through it and it's an oily, mushy sound. And you record that and build it up in several layers and you can make a nice sense of slimy snakes moving around the room.[6]

⌐∿

Some years ago, I discovered a way to produce a large number of *phantom words* and phrases within a short period of time.[7] Sit in front of two loudspeakers, with one to your left and the other to your right. A sequence is played that consists of two words, or a single word that's composed of two syllables, and these are repeated over and over again. The same repeating sequence is presented via both loudspeakers, but offset in time so that when the first sound (word or syllable) is coming from the speaker on your left, the second sound is coming from the speaker on your right, and vice versa. Because the signals are mixed in the air before they reach your ears, you are given a palette of sounds from which to chose, and so can create in your mind many different combinations of sounds.

A number of "phantom word" sequences are posted in the "Phantom Words" module. On listening to one of these sequences, people initially hear a jumble of meaningless sounds, but after a while distinct words and phrases appear. Those coming from the speaker on the right often seem to be different from those coming from the speaker on the left. Then later, new words and phrases appear. In addition, people often hear a third stream of words or phrases, apparently coming from some location between the speakers. Nonsense words, and musical, often rhythmic sounds, such as percussive sounds or tones, sometimes appear to be mixed in with meaningful words. People often report hearing speech in strange or "foreign" accents—presumably they are perceptually organizing the sounds into words and phrases that are meaningful to them, even though the words appear distorted in consequence. To give an example of the variety of words that people hear when listening to these sounds, here are some reports from students in a class I teach at the University of California, San Diego, when I presented *nowhere.*

*window, welcome, love me, run away, no brain, rainbow, raincoat, bueno, nombre, when oh when, mango, window pane, Broadway, Reno, melting, Rogaine.*

Shawn Carlson, in a *Scientific American* article titled "Dissecting the Brain with Sound," gave a vivid description of his experience when he visited my lab and heard the repeating words *high* and *low* alternating between two loudspeakers:

> The result is a pattern that sounds like language, but the words are not quite recognizable.
>
> Within a few seconds of listening to this strange cacophony, my brain started imposing a shifting order over the chaos as I began hearing distinct words and phrases. First came, "blank, blank, blank." Then "time, time, time." Then "no time," "long pine" and "any time." I was then astonished to hear a man's voice spilling out of the right speaker only. In a distinct Australian accent, it said, "Take me, take me, take me!"[8]

If English is your second language, you may hear some words and phrases in your native language. The students at our university are linguistically very diverse, and those in my class, on listening to these effects, have reported hearing words and phrases in Spanish, Mandarin, Cantonese, Korean, Japanese, Vietnamese, Tagalog, French, German, Italian, Hebrew, and Russian—to name just a few.

**Phantom Words**

http://dianadeutsch.net/ch07ex01

It's not unusual for the students to feel strongly that such "foreign" words have been inserted into the tracks, and they sometimes adamantly stick to this belief, despite my assurances to the contrary. Once, after I had played one of the "phantom words" to a large class, an exchange student from Germany raised her hand and called out, "You inserted 'genug' into the track." (*Genug* means *enough* in German).

"But I didn't," I replied, "The track consisted of only one word that was repeated over and over."

"But you definitely inserted 'genug,'" she insisted. "I know, because I heard it clearly."

"But I didn't."

"Yes, you did."

This went on for a while, and finally, because time was running out, I turned to a different subject. But when the class was over, the student approached me and announced: "I *know* you inserted 'genug' into the track." Then she turned and walked away.

People often seem to hear words and phrases that reflect what's on their minds—rather as in a Rorschach test,[9] though it's my impression that the present effect is stronger. I can bet who is likely to be on a diet, as they report hearing phrases such as "I'm hungry," "Diet Coke," or "feel fat." And students who are stressed tend to report words that are related to stress. If I play these demonstrations close to exam time, students may well hear phrases like "I'm tired," "no brain," or "no time." And interestingly, female students often report the word "love," while male students are more likely to report sexually explicit words and phrases (or sexually implicit—one male student, when presented with the sequence composed of *high* and *low* repeatedly heard the phrase "Give it to me"). So these illusions show that when people believe that they are hearing meaningful messages from the outside world, their brains are actively reconstructing sounds that make sense to them.

On May 11, 2018, Katie Hetzel, a 15-year old freshman at Flowery High School in Georgia, discovered a bizarre phenomenon. She was looking up the word "laurel" for a class project, and played an audio recording of this word from Vocabulary. com. But instead of hearing "Laurel" she heard the word "Yanny." Puzzled, she played the audio to her friends in class, and, as she put it, "we all heard mixed things." So she posted the audio clip to Instagram, and it was soon after posted as a video, with the word playing repeatedly, on Reddit, Twitter, Facebook, and YouTube.

From then on, the postings went viral. The Internet was torn apart for days, divided into two camps—those who heard the repeating word as "Laurel," and those who heard it as "Yanny"—with even married couples fighting over it. Notable people who ventured strong opinions included Ellen DeGeneres, Stephen King, and Yanni. Even President Donald Trump weighed in, saying "I hear covfefe"— referring to an earlier tweet of his. Various members of the White House staff also expressed themselves strongly, disagreeing completely among themselves as to which word was being spoken.

The Laurel versus Yanny effect is very similar to my "phantom words" illusion. One difference is that this effect is posted as a video displaying the words YANNY and LAUREL in large letters, with the word VOTE below them. Most people listen to the repeating sound while viewing the words, and so are primed to hear either one word or the other. (As discussed earlier, priming can have a strong effect on perception.) In contrast, in my phantom words demonstrations, I ask people to report in an open-ended way what they hear, and they often hear words that are related to what is on their minds.[10]

᧑

At all events, the voices that are "heard" in the phantom words demonstrations are remarkably similar to those reported by people who research "electronic voice phenomena" (EVP). These are sometimes taken as evidence of contact with the spirit world.[11] The idea that people can communicate with spirits of the dead goes back to ancient times, and exists in many primitive cultures. In the United States, a hugely popular spiritualist movement appeared in the mid-nineteenth century. This was largely centered on the Fox sisters from Hydesville, New York, who claimed that they communicated with the dead. Despite being constantly denounced as frauds, these sisters enjoyed enormous popularity, drawing crowds in the thousands. They and many others who joined the spiritualist movement held séances in which disembodied voices appeared, tables levitated, objects flew around the room, and glowing spirits were seen to materialize.

By the end of the nineteenth century, this spiritualist movement had largely petered out, and interest in EVP really began with the work of the artist Friedrich Jürgenson in the mid-twentieth century. Jürgenson was recording bird songs on a tape recorder, when he "heard" human voices on his tapes. Intrigued, he made many further recordings, and continued to "hear" unexplained voices, including some utterances that he believed were coming from his deceased mother. From these experiences, he concluded that the voices were emanating from spirits in the afterlife.

Jürgenson's work in turn inspired the parapsychologist Konstantīns Raudive to make many thousands of recordings, including some from a radio that was not tuned to any particular station, so that it produced a haze of static or noise. From these recordings he "heard" voices that were often distorted, and spoke with foreign accents or in different languages—often switching abruptly from one language to another—and often in a definite rhythm. Furthermore, some of the phrases he "heard" appeared to be related to his own personal experiences. Although there are other possible explanations for the phenomena he described—including interference from nearby CB radios, or even outright forgery—the characteristics of these voices bear a striking resemblance to the "phantom words" that I created, so it seems that a similar perceptual process is at work here. A strong believer in parapsychology might argue that I had stumbled upon a way of invoking voices from the spirit world, but then they would need to explain why different people hear different words and phrases when listening to identical recordings!

The 2005 movie *White Noise* takes you into the eerie and seductive world of EVP. Architect Jonathan Rivers, distraught over the accidental death of his wife, is persuaded by the EVP enthusiast Raymond Price to attempt to contact her. The audience is introduced into Price's studio, which is stocked with electronic and other paraphernalia—computers, video monitors, loudspeakers, headphones, microphones, TVs, radios, reel-to-reel, cassette, digital audiotape and videocassette recorders—along with books, tapes, and files scattered about. Seduced by the faint images and sounds that he hears, Jonathan turns into an EVP zombie, staring for hours into TV monitors displaying visual static, or listening to the audio static of detuned radios.

In real life, people mourning the death of their loved ones can be seduced all too easily into believing that they have made contact with them, particularly as bereaved people sometimes hallucinate the voices of their deceased spouses.[12] Michael Shermer, founding publisher of *Skeptic* magazine, described in *Scientific American* an interview with the paranormal investigator Christopher Moon, president of *Haunted Times* magazine, who claimed that he was able to make successful contact via a machine he called "Telephone to the Dead." (This machine was apparently derived from one claimed to have been invented by Thomas Edison—a claim that hasn't been validated.) The device rapidly tunes and retunes an AM receiver, then amplifies the resultant audio and feeds it into an echo chamber where (it is claimed) spirits further manipulate it so that their voices can be heard. However, Moon was unable to make contact with spirits in the afterlife during his interview with Shermer.[13]

The use of ambiguous visual patterns to evoke percepts that make sense to us is well known. For example, we see faces, grotesque figures, and landscapes in the clouds. Photographs of galaxies and nebulae from the Hubble Telescope provide particularly compelling examples of such misperceptions. Some of these are distinctly ambiguous, so that in gazing at these images your perception constantly changes—now you see an animal, now a bird, now a human face (human faces are particularly common), now a mountain scene, now a group of people walking—the image can appear in a constant state of flux. Other images bear such a striking resemblance to a familiar object that we perceive only this object and none other. The image of the Helix Nebula in Figure 7.1 looks so much like an eye that it is difficult to see it any other way—it has been dubbed "The Eye of God"—even though in reality it is a trillion-mile long tunnel of glowing gases.[14]

Ambiguous figures are useful for inspiring creativity. The eleventh-century Chinese artist Sung Ti gave the following advice:

You should choose an old tumbledown wall and throw over it a piece of white silk. Then, morning and evening you should gaze at it until, at length, you can see the ruins through the silk, its prominences, its levels, its zig-zags, and its cleavages, storing them in your mind and fixing them in your eyes. Make the prominences your mountains, the lower part your water, the hollows your ravines, the cracks your streams, the lighter parts your nearest points, the darker

FIGURE 7.1. The Helix Nebula, referred to as "The Eye of God." It looks strikingly like an eye, but in reality it is a trillion-mile-long tunnel of glowing gases.

parts your more distant points. Get all these thoroughly into you, and soon you will see men, birds, plants, and trees, flying and moving among them.[15]

In a similar vein, Leonardo da Vinci wrote in his *Treatise on Painting*:

You should look at certain walls stained with damp, or at stones of uneven color. If you have to invent some backgrounds you will be able to see in these the likeness of divine landscapes, adorned with mountains, ruins, rocks, woods, great plains, hills and valleys in great variety: and then again you will see there battles and strange figures in violent action, expressions of faces and clothes and an infinity of things which you will be able to reduce to their complete and proper forms. In such walls the same thing happens as in the sound of bells, in whose stroke you may find every named word which you can imagine.[16]

The psychologist B. F. Skinner created a method in the 1930s to generate such misperceptions from ambiguous sounds, using an algorithm that he called the "verbal summator." He recorded faint and indistinct vowel sounds in haphazard order, except that they were arranged in sequences approximating those of the common stress patterns in everyday English speech. He played these recordings repeatedly to subjects, and they reported hearing meaningful utterances—sometimes even phrases that appeared to refer to themselves.

In a related experiment, the Dutch psychologists Harald Merckelbach and Vincent van de Ven played students a recording of Bing Crosby singing "White Christmas." The students then listened for three minutes to a tape playing white noise, having been told that this song might be faintly embedded in the noise. A surprising thirty-two percent of the students reported that they heard the song, even though it had not been included on the tape.[17]

A different method for producing verbal transformations, termed *verbal satiation*, was studied in the early part of the twentieth century, and in the 1960s there developed a resurgence of interest in the procedure, particularly with the work of the psychologist Richard Warren. Here a single word is repeated over and over again, so that its meaning appears to degrade, and the listener generates different meanings instead. Describing this procedure, Warren wrote that this "satiation" effect works best when the word is played loudly and clearly, rather than indistinctly.[18]

⌒

That word-like nonsense sounds can be interpreted convincingly as meaningful words lies behind the concept of *backmasking*. This term is derived from the phrase "backward masking," which is employed by psychoacousticians to indicate something entirely different.[19] Here it refers to a message that is recorded backward onto an audio track and then played forward. The first well-known example occurs in the Beatles song "Rain." Apparently in the production process, John Lennon accidentally played a segment of the song backward, and the group liked the effect so much that they used it. The later Beatles song "I'm So Tired" ends with a phrase that sounds like gibberish when played forward, but when played backward some people hear it saying "Paul is a dead man, miss him, miss him, miss him." This provided fuel for an urban legend that Paul McCartney had died and had been replaced by a lookalike and soundalike—a legend that was also fueled by another Beatles song, "Revolution 9," which begins with a voice repeatedly saying "number nine" and when played in reverse, is heard by some people as "Turn me on, dead man."

Rumors concerning backmasking created a frenzy of activity. Some rock groups deliberately inserted reversed phrases into their music, and fans hoping to discover satanic messages played records backward with obsessive zeal. Equally zealous evangelists such as Pastor Gary Greenwald denounced such recordings as the work of Satan, and held huge rallies on the evils of rock music, which sometimes culminated in mass record-smashing events. In 1983, a bill was introduced to the California State Senate mandating that a record containing backmasking include a label warning that the masked message might be perceptible at a subliminal level when played forward.

Prompted by an inquiry from a local radio announcer, the psychologists John Vokey and Don Read at the University of Lethbridge carried out a series of experiments on the perceptibility of backmasked phrases.[20] In one test they created sentences, some of which were meaningful and others nonsensical. They played these sentences backward, and asked listeners to judge for each sentence whether it would make sense if heard in the forward direction—and they found that the listeners' performance was slightly below chance. In another experiment, they asked listeners to sort backward statements into one of five content categories: nursery rhymes, pornographic, Christian, satanic, and advertising—and again their performance was no better than chance. From these and other experiments, the psychologists concluded that there was no evidence that people are influenced by meaning in backward speech; rather, when people believe that they "hear" meaningful messages, their brains are actively constructing sounds that make sense to them, based on their knowledge and expectations.

There is another example of the brain reconstructing meaningful speech from ambiguous signals. This is known as *sine-wave speech*, and was first generated by the psychologist Robert Remez and his colleagues.[21] Three or four time-varying sine waves are synthesized, and these track the formant frequencies of a naturally spoken sentence. (For a detailed explanation of sine waves, harmonics, and formant frequencies, see physicist Eric Heller's book *Why You Hear What You Hear*.[22]) When this pattern is played out of context, it sounds like a number of whistles. But when the original sentence is played beforehand, so that the listener knows what to listen for, the brain reconstructs the sounds with the result that the same pattern is perceived as a spoken sentence.

But how, in the brain, can such reconstruction occur? According to the classical view, when a sound is presented, the signal is transformed by the inner ear (the cochlea) into electrical signals that travel up a chain of neuronal structures that constitute the auditory pathway. At each structure along the chain, neurons send signals to structures further up the chain, and ultimately to the auditory cortex and association cortex, where the sounds are interpreted. However, the process is much more complicated than this: Neurons in the auditory cortex send signals to other neurons in the auditory pathway, including those at lower levels, which in turn send signals to neurons even lower down, and so on, all the way back to the cochlea. Neurons in structures on each side of the brain also interact with each other. So the final interpretation of sounds from the cochlea involves an elaborate feedback loop, in which signals are transformed many times. To complicate matters further, auditory neurons also receive signals from other brain structures that convey different types of sensory information, such as vision and touch. And even further, neurons in the auditory cortex are influenced by signals from brain regions that underlie processes other than perception, such as memory, attention, and emotion.

Because the auditory system is composed of a highly complex set of interconnections, the signals at each level of the auditory pathway are modified in specific ways—some aspects are enhanced and others are dampened—and these modifications are influenced by the listener's experience, expectations, and emotional state. As a result, the signals that ultimately determine conscious perception have literally been altered in the brain, and so could appear convincingly to the listener to have arrived from the outside world in an altered form.

The perception of speech can be strongly influenced by input from the visual system. A striking demonstration of this influence is known as the *McGurk effect*.[23] A video of a talking head is displayed, in which repeating utterances of the syllable *ba* are dubbed onto lip movements for the syllable *ga*. On viewing this video, people generally hear neither *ba* nor *ga*, but fuse the input from the two senses so as to hear *da* instead. Numerous videos of this effect have since been posted.

The McGurk effect illustrates the fact that when people have acquired lip reading—because they are somewhat deaf, or work in a noisy environment—they don't just infer what is being said from the visual information provided by movements of the lips, but the visual signal actually affects the sounds they hear, and can even cause sounds to be "heard" when they are not really present.

The poet and writer David Wright became profoundly deaf after contracting scarlet fever at age seven. Yet he didn't originally notice his deafness, since he had already been unconsciously translating visual motion into sound. As he wrote:

My mother spent most of the day beside me and I understood everything she said. Why not? Without knowing it I had been reading her mouth all my life. When she spoke I seemed to hear her voice. It was an illusion which persisted even after I knew it was an illusion. . . . One day I was talking with my cousin and he, in a moment of inspiration, covered his mouth with his hand as he spoke. Silence! Once and for all I understood that when I could not see I could not hear.[24]

⌒

Not only do we mishear words and phrases, but we also sometimes misspeak ourselves, and produce unintended utterances. The person who is most credited for such "slips of the tongue" was the Reverend William Archibald Spooner, a much beloved don at Oxford University who served as dean and then as warden of New College. His "spoonerisms" are famous. For example, in toasting Queen Victoria at a dinner he is rumored to have said, "Give three cheers for our queer old dean." In talking to a group of farmers, he apparently described them as "noble tons of soil." In berating a student, he is said to have declared, "You have hissed all my mystery lectures. In fact, you have tasted two whole worms and you must leave Oxford this afternoon by the next town drain."[25]

While slips of the tongue provide food for entertainment, they also shed light on the workings of our speech processing system. The psychologist Bernard Baars devised a clever way to analyze these slips experimentally.[26] His subjects viewed strings of word pairs that they read silently. They were also cued at times to read certain "target" pairs aloud. "Interference" pairs, designed to produce spoonerisms, preceded the target pairs. For example, the subject might silently read the pairs "barred dorm" and "bought dog" just before seeing and reading aloud the word pair "darn bore." Under these conditions, many subjects produced the spoonerism "barn door." Baars and his colleagues found that their subjects were more likely to generate slips of the tongue that were real words rather than nonsense; they also tended to produce grammatically correct and meaningful phrases. So it seems that an "editor" in our verbal processing system at some point cleans up our utterances before we produce them.

In his book *Um . . . ,*[27] Michael Erard suggests an informal experiment that shows how well "priming" with other words can cause you to induce subjects to misspeak themselves. Ask a friend to say the word "poke" seven times. Then ask her what she calls the white of an egg. It's likely that your friend will reply "yolk." Try it—it works well, and shows that the "priming" process stimulates the neural circuits involved in the "oke" sound, and so interferes with the correct answer.

᳅

So far, we've been focusing on illusions derived from sounds of music and speech that have arisen in the real world. The next two chapters—on earworms and on hallucinations of music and speech—show that our brains can create sounds even in the absence of an outside stimulus. These sounds appear either as intrusive thoughts, or as having originated from the outside world.

# 8

## Catchy Music and Earworms

Music is so naturally united with us that we cannot be free from it even if we so desired.
—BOETHIUS, *De Institutione Musica*

A WHILE AGO, when I was shopping at the supermarket, I suddenly heard, in my mind's ear, a popular jingle advertising a hot dog. The jingle begins:

> Oh, I wish I were an Oscar Mayer Wiener
> That is what I'd truly like to be

As soon as the jingle came to an end it looped around and started all over again—and it went on looping in this fashion, plaguing me constantly for the next few days.

I had developed a bad case of *stuck tune*, or *earworm*—a peculiar malady that strikes most people from time to time. A tune or other musical fragment bores deep into our heads and plays itself until it arrives at a certain point, and it then proceeds to replay itself over and over, looping sometimes for hours, days, or even weeks. With a little effort, we can replace the tune with a different one, but then the new tune loops repeatedly inside our heads instead.

Many people find earworms quite tolerable. But for others, these persistent experiences interfere with sleep, interrupt thought processes, and even impede the ability to conduct a coherent conversation. A few people end up in doctors' offices in vain attempts to get rid of these remorseless intruders.

Descriptions of offensive earworms abound on the Web, along with suggested attempts to cure them. Some recommended ways of ousting earworms include chewing on cinnamon sticks, taking a bath, doing long division problems, solving anagrams, and replacing the objectionable tune with a different one. Favorite earworm replacements include "God Save the Queen,""Happy Birthday," "Greensleeves," and Bach fugues. One correspondent wrote to me that he blocks offensive earworms by counting numbers in series. Another correspondent obtains relief by writing down the words of the intruding song repeatedly in different orderings. My own favorite earworm-ousting technique is to ask myself why I'm experiencing this particular song, and this generally causes the stuck tune to disappear. But one thing that doesn't work is to try consciously to suppress the tune from my thoughts. This is like being asked not to think about pink elephants—as soon as someone tells me not to do so, I find myself compelled to imagine them.

Earworms have inspired a surprising amount of research. James Kellaris conducted a large-scale survey of earworm episodes.[1] He found that earworms were pervasive, with 98 percent of the respondents having experienced them. Most of the respondents suffered earworms frequently, with episodes lasting on average for a few hours. Songs with lyrics became stuck most often, followed by commercial jingles, and then by instrumental tunes without words. Musical features that were conducive to developing earworms included repetitiveness, simplicity, and some measure of incongruity.

Victoria Williamson and her colleagues at Goldsmith's College in London later carried out a study together with BBC Radio 6 Music.[2] Listeners to the radio program were invited to describe their earworms, and the researchers conducted an online survey to explore the conditions from which they arose. They concluded that earworms consisted mostly of familiar and recently heard music. They also noted that various events in the environment could trigger earworms—such as seeing a particular individual or hearing him mentioned, or hearing a certain phrase such as the title of a song. Recalling a meaningful event could also trigger an earworm.

The tendency to play out tunes in our heads, once we have begun to imagine them, appears to be built into our musical brain circuitry. In one study,[3] fMRI[4] was used to explore neural activity during listening to excerpts of popular music. The researchers first asked subjects to listen to excerpts of songs (such as "Satisfaction" by the Rolling Stones) and instrumental music that contained no lyrics (such as the theme from *The Pink Panther*), and to rate them for familiarity. They then extracted short sections (of 2–5 sec duration) from the soundtracks, and replaced these with silence. Remarkably, the fMRI scans showed more activation in the auditory cortex and association cortex during silent gaps that were embedded in the familiar music, compared with music that was unfamiliar. And correspondingly, although no instructions were given to the subjects, they all reported that they heard the familiar music continue through the silent gaps. So people have a tendency to continue "hearing" a piece of familiar music just after they have heard part of it. Although this doesn't explain why earworms keep looping in our heads repeatedly, it's probably part of the picture.

We are experiencing an epidemic of stuck tunes, and its cause isn't hard to find—we are deluged with music through most of our waking lives. Background music is heard everywhere—in restaurants, department stores and supermarkets, in hotel lobbies, elevators, airports, railway stations, libraries, doctors' waiting rooms, and even hospitals. In addition, many people listen to music fairly continuously over their radios and televisions, and on stereo systems, computers, and portable media players such as smartphones. This constant exposure to music could sensitize our musical processing systems so strongly that they tend to fire off spontaneously.

While many people like background music, its ubiquity has produced a backlash among those who find it aggravating, and this has led to the formation of protest groups. The largest organization that lobbies for the abolition of background music in public places is Pipedown,[5] which reports thousands of paying members who distribute preprinted protest cards in their campaign to restore freedom from unwanted music. They point to a survey of over 68 thousand people, that was conducted by the management of London's Gatwick Airport, in which 43 percent of the respondents said that they disliked piped music, while only 34 percent said that they liked it—and in consequence Gatwick discontinued having music piped into their main areas. The *Sunday Times* conducted a poll asking readers what was "the single thing they most detested about modern life," and piped music came third on the list—behind two other forms of noise. A BBC poll of commuter train travelers from Essex to London found that 67 percent of the respondents objected to having piped television in the trains—some even barricading themselves in the toilets to avoid the unwanted sound. As a result, this train service discontinued piped television, and the management even created Quiet Zones in which people were asked not to use their cell phones or other noisy electronic devices. Pipedown also helped persuade Marks & Spencer, the biggest chain store in the United Kingdom, to drop piped music.

A UK National Opinion Poll of people with hearing problems found that 86 percent of the respondents hated piped music. This is not surprising, since hearing-impaired people are particularly vulnerable to noise when they are trying to understand speech. As another serious issue, blind people find background music disorienting, since they rely on sound cues to help navigate their environment, such as when they are crossing a busy street.

My most memorable encounter with piped music occurred when, following a meeting of the Acoustical Society of America in Hawaii, my husband and I decided to visit Lahaina, on the island of Maui. After a short airplane ride, we rented a car and drove past sunny fields of sugar cane until we arrived at our destination. We were greeted in the hotel lobby with piped music. The hotel restaurant had piped music, and in an attempt to escape it we explored other restaurants in the area— but they all had piped music. So we thought that surely if we took a boat tour out to sea we would escape it—but on enquiry we found that music was being piped into the boats. Finally, having failed on land and on sea we enquired about heli-copter tours. "Certainly we have piped music in our helicopters" the agent proudly declared. So having failed on land, sea, and air, we gave up, and drove to a remote part of Maui for the rest of our vacation.

Many musicians object strenuously to piped music—probably because they feel impelled to listen to it. I'm unable to think coherently about anything else while music is playing—whether I like the music or not, my attention swoops down on it, and largely blocks out other thoughts. Musicians who have spoken out against piped music include the conductor Simon Rattle, the cellist Julian Lloyd Webber, and the pianist and conductor Daniel Barenboim, who delivered an angry attack on this violation of his personal space. In his 2006 Reith Lecture in Chicago,[6] Barenboim declared:

> I have been on more than one occasion subject to having to hear, because I cannot shut my ears, the Brahms violin concerto in the lift, having to con-duct it in the evening. And I ask myself, why? This is not going to bring one more person into the concert hall, and it is not only counter-productive but I think if we are allowed an old term to speak of musical ethics, it is absolutely offensive.

A further protest against the ubiquity of piped music occurred with the 1995 establishment of No Music Day.[7] This was the brainchild of former rock star Bill Drummond, co-founder of the highly successful band The KLF, which abruptly stopped performing at the height of its success. Drummond picked November 21 to observe this day, since November 22 is the feast of St. Cecilia, the patron saint of music. Drummond publically announced that on this day, "No records will be played on the radio," "Rock bands will not rock," "Jingles will not jangle," and "Milkmen will not whistle," among other prohibitions. In consequence, thousands of people in the United Kingdom pledged themselves not to play or listen to music on this day. BBC Radio Scotland decided in 1997 to observe No Music Day, and to devote radio time on that day to discussions about music instead.

ᴑ

So how did this state of affairs arise? Consider what the world was like before the end of the nineteenth century. There was no radio or TV; there were no phonographs, no cassette players, no CD players or iPods. Only live music was played or sung, and—apart from music lessons and people humming or whistling to themselves—music was heard only intermittently, and in certain venues such as churches, concert halls, and dance halls, or at special events such as birthday and wedding celebrations.

But all this changed with the development of recording technology, which enabled sounds to be captured and preserved on physical media, so that they could be heard anywhere at any time. Today, the ready availability of downloaded music, and free access to music over the Web, further contributes to the huge quantity of music to which we are exposed.

In fact, music is now so pervasive that it takes an effort to imagine what the soundscape was like a hundred years ago. When Thomas Edison invented the phonograph in 1877, he didn't expect it to be used primarily for music. Rather, he proposed that it be employed largely for language-based activities, such as dictation without a stenographer, and elocution lessons. It was only after the phonograph had been patented that entrepreneurs recognized the enormous commercial potential of recorded music. Once this was acknowledged, the music industry developed rapidly. Beginning in the late nineteenth century, composers, lyricists, arrangers, publishers, and promoters worked together with the intention of producing songs that would have large sales. This new industry, which was dubbed Tin Pan Alley, was centered in New York, and involved composers such as Irving Berlin, George Gershwin, and Cole Porter. Songwriting guidebooks advised hopeful composers to make their songs simple and familiar—qualities, as it turns out, that make music not just catchy but also sticky. In 1920, Irving Berlin proposed a set of rules for writing a successful popular song. Among these rules were:

> The title, which must be simple and easily remembered, must be *"planted"* effectively in the song. It must be emphasized, accented again and again, throughout the verses and chorus . . .
>     Your song must be perfectly *simple* . . .[8]

Elsewhere Berlin emphasized that a new song should contain familiar elements, writing, "There is no such thing as a new melody," and arguing that effective songwriters "connect the old phrases in a new way, so that they will sound like a new tune."[9]

As recording technology advanced, frequent repetition of phrases became even more common in popular music. In the 1970s, hip-hop DJs developed the technique of repeating a fragment of music indefinitely by switching between two turntables that contained copies of the same record. Loops then became a basic element of instrumental accompaniment in rap music. Later, the development of digital sound recording enabled loops to be created using sampling techniques, with the result that they have become key elements of most popular music today. In parallel, the same technological advances enabled the development of the minimalist school of composition, as represented by the composers Steve Reich, Philip Glass, and Terry Riley.

Of course, we can't attribute the repetitiveness of popular music entirely to technological advances. As the music theorist Elizabeth Margulis explains in her book *On Repeat*,[10] popular music must have tapped into a human appetite for repetition in music. In fact, ethnomusicologists have been arguing for decades that repetition is a fundamental characteristic of music in all cultures. Following this line of reasoning, the tendency to crave repetition in music could also be inducing us to replay musical segments in our imagination, so as to give rise to stuck tunes.

Why should repetition be such an important element in music? One part of the explanation is the *mere exposure effect*, which was studied by the psychologist Robert Zajonc in the 1960s.[11] Zajonc found that the more often we encounter a stimulus, the more positive our attitude toward it becomes. This holds for nonsense words, Chinese characters, paintings, and photographs, among other things. And it holds for music also—the more often we encounter a piece of music, the greater the probability that we will like it. The increased popularity of the piece then leads to its being played more frequently, and this in turn further increases its popularity—and so on, in a positive feedback loop.

The mere exposure effect is an example of *processing fluency*,[12] a term that indicates the ease with which we process a stimulus—for example, how quickly and easily we recognize it—and this in turn affects how pleasurable it is. This holds for many different objects and events. For example, we tend to prefer visual patterns when we see them more clearly, when they are presented for a longer time, and when we view them repeatedly.

The importance of familiarity to emotional reactions to music is supported by a study in which subjects rated their familiarity with song excerpts from the pop/rock repertoire, and also rated how much they liked the songs.[13] This enabled the

researchers to select songs that were personalized for each subject, both for familiarity and also for how much they liked them. The subjects then listened to the
musical excerpts while their brains were scanned using fMRI. The scans showed
more activation in response to familiar than to unfamiliar music in brain regions
involved in emotion and reward—and this was true regardless of whether or not
the subjects liked the music they heard. So this study showed that familiarity has
a strong effect on emotional engagement with music.

Margulis's research demonstrates the remarkable power of repetition on both
untrained and sophisticated listeners. She played excerpts of music by famous
20th-century composers such as Luciano Berio and Elliott Carter, who had deliberately avoided repetition in their compositions. Listeners without special musical
training rated how much they enjoyed each excerpt, how interesting they thought
it was, and how likely it was to have been composed by a human artist rather than
generated by computer.

In some of the excerpts, Margulis extracted segments of the music and rein-
serted them in other places so as to produce repetitions, without regard for artistic
quality. Remarkably, the listeners rated the excerpts containing repeated segments
as more enjoyable, more interesting, and even more likely to have been composed
by a human artist rather than a computer. Even more astonishing, when Margulis
played these excerpts to roomfuls of professional musicians at a meeting of the
Society for Music Theory, the members of the audience also preferred the versions
containing repetitions.

Repetition sells. Most of the popular music that people hear in their daily
lives involves lyrics. Joseph Nunes at the University of Southern California and
his colleagues carried out an analysis using Billboard's Hot 100 singles chart.[14]
This is the standard record chart for singles in the US music industry, and ranks
music based on sales, radio play, and online streaming. It is widely regarded as
the best source for assessing a song's popularity or "bigness of hits." Taking
a sample of 2,480 songs on the chart from its inception in 1958 to the end of
2012, the researchers identified all those that reached #1 (which they called
"Top Songs") and all those that never climbed above #90 on the chart (called
"Bottom Songs").

The findings concerning repetition were stunning: As the number of repetitions
of the chorus of a song increased, the probability that the song would be a Top
Song as opposed to a Bottom Song also increased. Further, as the number of repetitions of individual words in a song increased, the probability of the song be-
coming a Top Song as opposed to a Bottom Song increased. And focusing only on
the Top Songs, the researchers found that the greater the number of repetitions of
the chorus, the more rapidly the song rose to #1 on the chart.

The effect of fluency alone doesn't account for our strong appetite for repeating music, however, since the fluency effect holds true for exposure to other things as well—things that don't compel our attention in the same way. We may view favorite photographs or paintings many times, but we don't feel impelled to view them constantly, as can happen when we are gripped by a piece of music. Once we have listened to part of a familiar musical phrase, we can become driven to play it out in our heads. As Margulis wrote:

> One reason for this stickiness is our inability to conjure up one musical moment and leave it; if our brain flits over any part of the music, we are captured by it, and must play it forth to a point of rest. So we constantly have a sense of being gripped, even unwillingly, by the tune. (p. 11)

So while we understand some of the characteristics of music and brain function that are associated with stuck tunes, there is much that remains mysterious about this bizarre phenomenon. It's also paradoxical that while repeating a segment of music generally produces a positive effect, many people object strongly to being bombarded with background music, and the emotional response to stuck tunes is often very negative.

As eloquently described by Oliver Sacks in his book *Musicophilia*,[15] memories of events that we experienced long ago can have enormous power to evoke musical images, including stuck tunes. Recollections of my childhood in London are always associated with music. My parents were not musically trained, but they had a passionate love of music, and our house was filled with song. I remember many evenings around the dinner table when my father would suddenly burst into song, and my mother and I would join in. My mother had a beautiful contralto voice, and some of my earliest memories are of her singing folk songs as she went about her work. Many of the songs I heard at that time are deeply etched in my memory, and it takes very little for one of them to pop into my head and play itself repeatedly. These stuck tunes are always welcome, and imbued with nostalgia.

The power that earworms have to take over our minds has led to their effective use in movies as devices to help the narrative along. Alfred Hitchcock, who likely experienced stuck tunes himself, used them as central themes in two of his great movies. In *The 39 Steps*, the hero, Richard Hannay, hears a tune in a music hall, and is unable to get it out of his head, so he hums and whistles it to himself obsessively. At the end of the movie, the significance of the tune is revealed to the hero, so that he is finally able to act effectively.

In another Hitchock movie, *Shadow of a Doubt*, the heroine, Charlie, is plagued by misgivings about her visiting uncle. These misgivings emerge when the "Merry

Widow Waltz" begins to run through her mind. They turn out to be justified when her uncle is revealed to be the "Merry Widow Murderer," who is wanted by the police for strangling three wealthy widows to death. Not only does the waltz haunt Charlie, but it's also played in distorted forms throughout the movie, causing the tune to be stuck in our minds as well, and so to remind us constantly of the villain's sinister identity. In both movies, Hitchcock interprets the stuck tunes as conveying messages that leak out from the unconscious mind.

❧

Spoken phrases don't generally get stuck in our heads as do fragments of music. On the other hand, it's possible that spoken jingles and doggerels turned into earworms more frequently in the days before music became so pervasive. In 1876, an incident occurred that formed the basis of a famous short story by Mark Twain.[16] Two friends, Isaac Bromley and Noah Brooks, while riding a tram, had noticed a sign that informed passengers about the fare. Intrigued, they used it as the basis of a jingle that they published together with their friends W. C. Wyckoff and Moses Hanley. The jingle quickly achieved widespread popularity, and inspired Mark Twain to write his story, which he entitled "A Literary Nightmare." The jingle goes:

> Conductor, when you receive a fare,
> Punch in the presence of the passenjare!
> A blue trip slip for an eight-cent fare,
> A buff trip slip for a six-cent fare,
> A pink trip slip for a three-cent fare,
> Punch in the presence of the passenjare!
>
> CHORUS.
>
> Punch, brothers! punch with care!
> Punch in the presence of the passenjare!

In this story, Twain describes how he came across the jingle in a newspaper. It instantly took entire possession of him and "went waltzing through [his] brain," so that he was unable to tell whether or not he had eaten his breakfast, his day's work was ruined, and he was unable to read, write, or even sleep. All he could do was rave "Punch! Oh, punch! Punch in the presence of the passenjare!" Two days later, he became rid of the jingle by unwittingly passing it along to a friend, who in turn was rendered incapacitated, and could only repeat the remorseless rhymes. His friend was finally saved from an insane asylum by discharging the persecuting jingle into the ears of a group of university students.[16]

❧

The tendency for music to get stuck in our heads makes it ideal for radio and TV advertisements—a catchy tune can turn into an earworm, and carry the name of the product along with it. But words are also very important to the success of ad jingles, and the fact that they are sung rather than spoken is a key factor here. People will shrug off a spoken message that is ridiculous or obviously manipulative. But if the same message is sung, we allow it to get through to us. For example, the "Oscar Mayer Wiener Song"—one of the most successful commercials ever—features children singing:

> Oh, I wish I were an Oscar Mayer Wiener
> That is what I'd truly like to be

But that's ridiculous—who would ever want to be an Oscar Mayer Wiener? The absurdity of this statement grabs your attention so that you listen through to the end of the advertisement, and hear the words 'Oscar Mayer Wiener" five times in 40 seconds! (Amazingly, the composer, Richard Trentlage, composed the jingle in an hour, after hearing his young son say, "I wish I could be a dirt bike hotdog." He took another 20 minutes to record his two children singing it in his living room—and the rest is history.)

A few years back, I asked Michael A. Levine, the composer of the enormously catchy and successful Kit Kat commercial "Gimme a Break" about the features he believes are responsible for the success of his jingle—beyond the obvious catchiness of the melody. (The jingle, which is often described as one of the catchiest of all time, is still going strong—and has been listed as one of the top ten earworms.) Levine described in a podcast of *Burnt Toast*[17] how he composed the jingle in an elevator ride to confer with the sponsors.

In his email back to me, Levine wrote:

Malcolm Gladwell articulated something that I felt but hadn't expressed as succinctly by saying that "stickiness" - an inherent property of earworms -is dependent on some level of "wrongness." If something is just the right amount out of whack - almost but not enough to attract conscious attention - then your subconscious goes over and over it trying to figure it out. I once referred to this kind of low-level wrongness as "a mimetic piece of corn stuck between your cognitive teeth" - a phrase Malcolm liked well enough to quote on his website.

This is true of both "high" and "low" culture. The emotional ambiguity of the Mona Lisa's smile. The rhythmic ambiguity of the opening of Beethoven's 5th (Triplets? Eighths?). The irony of Huck Finn debating whether it was ethical to help an escaped slave. "Shall I compare thee to a summer's day" in which Shakespeare spends the entire sonnet describing what his love is not like. The use of the colloquial "like" instead of "as" in "Winston Tastes Good Like A Cigarette Should" which offended 50s grammarians.

The Kit Kat jingle would have been a great deal less effective were it not for Ken Shuldman's "Gimme A Break" slogan. The play on words (Kit Kat bars are manufactured in pre-made sections designed to be broken off) coupled with the important fact that the expression "Gimme a break" had, previously, been considered a negative statement. So you have the bouncy little tune with an exasperated exhortation . . . that turns out to be a demand for a piece of candy! There is also a little bit of musical contradiction as well as the first four notes of the melodic line imply an ascending major pentatonic, followed by a descending line starting on the flat 7th - a "blue" note. So, not enough to consciously scream "Wrong!", but enough out of place to make your mind consider and reconsider what's going on.

Of course, all of that analysis was post hoc. I composed the music off the top of my head while riding in the elevator coming down from the meeting at the ad agency, holding two pages of Ken's ideas for possible lyrics in my hand, of which "Gimme A Break" and "Break me off a piece of that Kit Kat Bar" was all I used.[18]

As a logical consequence of the importance of familiarity to the "catchiness" of a tune, it makes sense to take a melody that has already been popularized, and set it to words that promote the product. Based on this line of reasoning, melodies that were already well known have been supplanting jingles in TV advertisements. In the 1980s, a TV commercial for Nike sneakers featured the Beatles song "Revolution." In the 1990s, an accounting firm used Bob Dylan's "The Times They Are a-Changin'" in a corporate image advertisement. And in the 1980s and '90s, the melody of the classic Neapolitan song "O Sole Mio" was set to different words ("Just One Cornetto / Give it to me") and used in a TV commercial selling Walls ice cream.

Many people decry the practice of using famous music to sell unrelated products, on the grounds that it debases the original music. Barenboim gave a particularly egregious example in his 2006 Reith lecture:

And the most extraordinary example of offensive usage of music, because it underlines some kind of association which I fail to recognise, was shown to me one day when watching the television in Chicago and seeing a commercial of a company called American Standard. And it showed a plumber running very very fast in great agitation, opening the door to a toilet and showing why this company actually cleans the toilet better than other companies. And you know what music was played to that? . . .

The Lachrymose from Mozart's *Requiem*. Now ladies and gentlemen, I'm sorry, I'm probably immodest enough to think I have a sense of humour but I can't laugh at this.[6]

Yet regardless of the intentions of the composer, the more powerful and evocative a piece of music is, the more likely it is to be etched in our memories, and to carry along with it associated messages that influence our behavior, including the purchase of a particular brand of ice cream.

◌

So far we have been considering music that appears to us as intrusive thoughts. We next inquire into phantom music that is heard as though it had originated in the real world. The musical hallucinations we'll be exploring are not only intriguing in themselves, but they also shed light on our normal perception of music.

# 9

## Hallucinations of Music and Speech

I wouldn't hallucinate songs that I knew, I hallucinated sophisticated, big band jazz, multiple horn sections, fully harmonized, completed with a highly intricate rhythm section. I was hearing fully completed scores.

—WILLIAM B.

ONE DAY IN July 2000, when Barbara G. was purchasing some clothes in a department store, it suddenly appeared to her that the sung chorus of "Glory, Glory, Hallelujah" was being piped loudly and repeatedly through the sound system. After hearing about twenty repetitions of this chorus she felt she could stand it no longer, and left the building in exasperation. However, she was amazed to find that the chorus followed her into the street, singing just as loudly and repeatedly.

Alarmed, she returned to her apartment, only to find that "Glory, glory, hallelujah" returned with her. She closed all the windows, stuffed her ears with cotton wool, held pillows against her head, and still finding no relief, consulted an audiologist.

Barbara was elderly and had hearing problems, but appeared quite normal on psychiatric and neurological examination. Her audiologist, knowing of my interest in unusual aspects of music perception, suggested that she contact me. When I spoke with Barbara, her hallucinations had been continuing incessantly for months.

Around this time, I received a telephone call from Emily C., an articulate lady who was also somewhat hard of hearing. Similarly to Barbara, her musical hallucinations had begun dramatically: She had been awakened suddenly in the middle of the night by what appeared to be a band playing loud gospel music in her bedroom. Once she was fully awake she realized that, of course, there was no band in the room. Her next thought was that a clock radio must have started up, and was hidden somewhere. She rummaged in her drawers, hunted through her closets, searched under her bed, but found nothing. Exasperated, she walked out of the room, only to find that the music followed her into the corridor, and continued with her into all the rooms in her apartment.

Emily then concluded that the music must be playing in a different part of the building, so she went downstairs to complain to the doorman. "But I don't hear any music," he declared, whereupon Emily, now thoroughly exasperated, called the police. When they also insisted that they heard nothing, she reluctantly concluded that the music must be coming from inside her head. By the time she contacted me, her hallucinated music had been continuing for many months, with no respite, and was interfering with her ability to concentrate, or even to get a reasonable night's sleep. Again, neurological and psychiatric examinations had turned up nothing abnormal.

I happened to relay such stories to Roy Rivenburg of the *Los Angeles Times* when he was interviewing me about stuck tunes, and he mentioned my interest in musical hallucinations in an article that appeared in 2001. As a result, I received hundreds of emails from people who hallucinated music—several of whom wrote to say how relieved they were to know that this didn't mean they were going crazy. Many of my correspondents experienced dramatically intrusive, persistent, and unwanted hallucinations. But others instead heard phantom music that was quiet, beautiful, and sometimes original.

In particular, I was fortunate to have had a detailed correspondence with Hazel Menzies, a fine musician and a physicist, whose hallucinations were intermittent, complex, and beautiful. Her descriptions provide a remarkably informative account of the phantom music that she heard. Hazel first contacted me on March 2002. In an email entitled "Ghostly Choruses," she wrote:

> AT LAST some indication that I am not going mad!!!! I had truly come to believe that the "music" I sometimes hear might be an indication of some form of insanity and had not mentioned it to anyone for fear of being laughed at or thought to be mad. Silly, you'd probably think but, what can I say? That's how it made me feel, especially since it was religious-sounding and I've never had ambitions to be the next Joan of Arc.

I replied to this email by asking her some questions about the content of her hallucinations. She replied:

Err . . . I hadn't really thought about this in terms of specific questions, excuse me while I ramble . . .

". . . do you hear primarily male voices, or female voices?" Well, I hear all singing voice types, like a whole chorus. The music is dominated by the higher pitched voices, female/boy types, but in real choral music this is the case, simply because that's the way it is performed/written. So I couldn't say whether the dominance by the higher voices is because my brain might be creating it that way or because that's the way my experience tells me it *should* be.

Hazel continued answering my questions:

"Does the music tend to be religious in nature, or does it tend to have a different cast?" The music is very choral indeed. I think one might tend to associate it with religious music since religious music is generally pretty choral in form. I hear swells and dips in intensity and prominence. It's mostly choral but there's random identifiable single strands to it. Although it sounds choral, I don't actually hear lyrics, as such, just word-type sounds, mainly vowels but also some sibilants. The closest "real" music with which I could compare the style would be, say Vaughn Williams' Sea Symphony, the darker parts of some of the Requiems (Faure/Mozart ish). You know how boy chorister's voices suddenly sing out above the mass? There's occasional strands like that which blast through an odd clear note above the mass behind. Sometimes the whole "chorus" sings one combination of notes in unison for a long time—longer than one real breath would allow.

However, the "sounds" could also be compared to the modern electronic type of music, generally referred to as "ambient"—I think of this type as "Ghost in the Machine" music. Blimey this is difficult to describe . . . maybe an example would help:

Say I'm in the car doing a long journey. I might take some tapes along to combat the boredom. Say the tape comes to an end and I'm in town traffic so I don't get another one out to play. Then, say, I just drive for a while. After a period, the musical hallucinations might start. The first time this happened I thought the tape had not played to the end and that I was hearing something left over from previous recording but sounding mangled as if the tape had been near a magnetic source or chewed up in the player and there was only residual recording left. Bloody spooky, actually, when I think about it, finding that it wasn't the tape player or some harmonic resonances from the structure of the car or engine - but me!

Hazel then responded to a different question:

"Do you 'hear' familiar music, or does it appear original?"

It's original in the sense that it does not follow a melody or formal structure that I recognise but it is familiar with respect to the mix of the voices which in the main follow the common rules of harmony, often 4 or more parts. Sometimes the chords get diminished or augmented, like jazz harmonies, but mostly they're the sort of harmonies that would pass for standard classical choral stuff.

I am the type of musician that can replay familiar musical works, in their entirety, in my head anyway. I don't have difficulty hearing more than one tune at once, say, if I need to work out harmonies for singing. If I get a tune stuck in my head it's not just the singer's voice, it's the whole caboodle, and it's accurate (painfully accurate in the case of some of the tripe my brain seems to latch onto). It's so frequent that I have familiar or catchy tunes playing away in my head that I take this soundtrack to my life totally for granted - it's only annoying if it's something I don't like. For instance, at the moment I have several tracks for the motion picture *O Brother Where Art Thou* jostling for prominence, these last few days. I am aware of this but it doesn't drive me mad because I enjoyed the film and early US music is a particular interest of mine atm. In the back of my mind I'm choosing how I would tackle some of the songs and deciding whether to play with any of them. I would call this "normal" and the musical hallucinations more "unusual."

"Does it appear to be coming from a particular position in space?" Now, I never thought about this! Yes, it seems to be located in my headspace, not external space. Contained within me, as it were, like my ears were hearing it from the opposite side of my eardrums to hearing external sounds. In fact, I have wondered whether I was hearing a physical resonance effect related to the change of velocity of blood as it travels around different sized veins or arteries in the inside of my ears. People speak of hearing the rushing of blood at times of stress/fear, don't they? I have wondered if what I hear was to do with that.

Hazel then went on to reply to another question:

"Do you welcome the music, or would you prefer not to hear it?" Well, as I have said, in a way it has frightened me somewhat. I am a person who prefers to have explanations for the various phenomena that we encounter in our lives. I'm a physicist by profession and therefore, to be experiencing something for which I cannot readily supply a feasible explanation is highly disconcerting.

Add to this the historical cultural abhorrence of things like hearing voices, witches, etc. and the modern fear and ignorance about mental illnesses such as schizophrenia and you can see that the appearance of these musical hallucinations coincident with me beginning to live alone for the first time (at 30y old) was . . . I'm searching for the right word, here . . . peculiar.

I asked Hazel whether she thought that since the music she "heard" was in some aspects original, she might, in a sense, have been composing it. Hazel wrote back:

No. I wouldn't say I was in any sense "composing" the music I hear since the hallucinations do not begin or end by any act of will on my part. If I try to impose my will whilst it happens, then the music turns into something I know and I go into "Stuck Tune Syndrome." However, I am not an experienced composer; at least, not in the formal, classical sense. Of course, I can write a tune, as can anyone!. . . . . .

A composer, if he were using musical hallucinations such as mine, would have to add formal structure to make it Music. It may be that a lot of stuff has been created that way. . . . Creative people will take their inspiration anyway they can. Bearing in mind that I was wary about mentioning the hallucinations to anyone for fear of being thought mad, would a composer not also be thinking that way? Therefore, any writing they might do on their thought processes would be guarded.

I also asked Hazel if she could notate the phantom music that she heard. Other musicians who hallucinated music had found this impossible, since it was too fleeting to be captured in notation—just as it is difficult to notate music that one hears in dreams. However, since Hazel's music tended to be slow, it seemed to me that she might be able to set some of it down. To my delight, on July 5, 2002, she sent me a couple of scans of the music she had hallucinated. (As she explained, the music had originally been fragmented, and she had juxtaposed the fragments in the scans.) Along with the notations, which are reproduced in Figure 9.1, she wrote:

I have used the musical notation (dynamics, timing, etc.) more to express the effect I hear rather than with the aim of giving someone the means to reproduce it . . . I am unsurprised to see that these musical hallucinations are in the key of Em, i.e., the open strings of the guitar, as I have been playing since I was 8 y old.

Hazel's descriptions of "hearing" beautiful and complex music were later joined by others. Some of these involved vivid hallucinations that occurred at the

FIGURE 9.1. Fragment of hallucinated music, heard and notated by Hazel Menzies. Reproduced by permission.

borderline between sleep and wakefulness (called "hypnogogic" when the person is falling asleep, and "hypnopompic" when the person is waking up). Such hallucinations are not uncommon. William B. described vivid and elaborate hallucinations that materialized in the middle of the night:

I wouldn't hallucinate songs that I knew, I hallucinated sophisticated, big band jazz, multiple horn sections, fully harmonized, completed with a highly intricate rhythm section. I was hearing fully completed scores. At the time,

I knew that it was different than the initial burst of inspiration that came with the usual songwriting process, which came in bits and pieces (e.g., fragments of melody or basic rhythmic pattern). This was different because it was like I was hearing a recording in my head, of the most wildest jazz fully intact—music that I'd never heard before. I didn't even try to get up and try to write it out because I knew I'd never remember anything but basic fragments, so I just laid there and listened to it, amazed at what I was hearing.

Colin C. gave a similar description:

so far my hypnagogic music has been all me—resplendent, complicated melodies, often brilliant, anarchic jazz, where each instrument's melody flies off independently, resonating briefly and unpredictably, meeting up, reintroducing some vague idea of harmony momentarily that could indicate that the whole song is simply one single extended, very complicated harmony. Other times it has been 80-piece orchestral magnificence. Very Straussian, Schubertian . . .

Music that's hallucinated on the threshold of sleep is probably related to music that's heard in dreams. On occasion while sleeping I've experienced extraordinarily beautiful music, as though sung by choirs with an orchestral accompaniment. For a while I would keep music manuscript paper by my bedside, in the hope of capturing some of the music before it disappeared. Yet it would vanish so rapidly that I could notate only one or two lines, and was never able to recreate the ethereal beauty of the sounds that I had heard while asleep.

A few composers—Beethoven, Berlioz, and Wagner among them—have described works that were derived from music they heard in dreams. The violinist/composer Guiseppe Tartini was inspired to write his most famous piece, the Violin Sonata in G Minor, by a dream in which he gave the Devil his violin. To his astonishment, the Devil played a piece that was so extraordinarily beautiful, and performed with such intelligence and artistry, that it surpassed anything he had ever heard or even imagined. Overcome by the music, Tartini woke up gasping for breath. Immediately he seized his violin in the hope of recapturing the music he had just heard, but it was useless. However, he then composed the piece known as the Devil's Trill Sonata, inspired by this experience. He considered it to be the best of all his works, yet he said that the music he had heard in his sleep was so superior to this composition that he would have smashed his violin and given up music forever if he could only have recaptured it.[1]

FIGURE 9.2. *Tartini's Dream* by Louis-Léopold Boilly. Illustration of the legend behind Giuseppe Tartini's "Devil's Trill Sonata." Bibliothèque Nationale de France, 1824.

But I don't want to give the impression that musical hallucinations are usually exciting and exhilarating. To the contrary, most people who've written to me about their hallucinations find them distressing, and are hoping for relief from the incessant loud music that plagues them. I asked Paul S. to describe in real time what he was hearing, and here's what he wrote:

> It's 9:41 PM here, and the music is playing loud. Now the music is locked into a repeating stanza of the "default song." There are numerous bagpipes playing the one stanza. There are also flutes that are cracked, and the ocarena. It is loud. I'm going to change the song. I can do this. I'll change the song to, "What Kind of Fool Am I."
>
> Now that is playing, but it's wretched, but the melody is definitive. This seemed to increase the volume level, and more instruments have joined in. All the same as the others. But now, there's a very high type of, well, bird warbling above all the other instruments. There's now a piercing high note— that of an instrument I can't identify. It could be a whistle of some sort. I'll change the song again, with a little association and concentration, as "What Kind of Fool Am I" has brought in too many high notes.
>
> I just changed the song to "Over the Rainbow." This is a little better. The melody is less wretched, but still way off-key. The instruments are the same,

but the ocarena is more pronounced. It's playing way faster than the original version. I can't slow it down . . .

More typically, people have described their hallucinated music in terms of "play-lists" that were composed of music in many different genres, and from which fragments were pulled out seemingly at random, as though from an iPod, a jukebox, or a radio that unpredictably switched channels. Harold M., a highly successful businessman, wrote:

Sometimes it's classical, sometimes pop, country or rock. I don't control the random play list nature of the song that is playing in the morning when I wake up . . . the tunes change all the time. Just today, as I was working out on my lunch hour, the songs were shuffling through my head like a musical sampler.

Overall, the fragments that people have hallucinated tended to consist of music they had known in childhood, such as patriotic songs, religious hymns, Christmas carols, folk songs, music from old TV ads and shows—though these were also mixed with other music that they'd heard later in life. One "playlist," sent to me by Jerry F., consisted of "Beautiful Dreamer," "Old Man River," "Tom Dooley," "Silent Night," "America the Beautiful," "Old Folks at Home," "Marines' Hymn," "Spring Fever," "Come All Ye Faithful," "Clementine," "Happy Birthday," "Old MacDonald Had a Farm," and "Auld Lang Syne." Jerry further commented: "I even heard a room full of small children singing the Star-Spangled Banner to an out of tune piano" Another playlist, sent by Brian S., included "Go Tell It on the Mountain," "Auld Lang Syne," "Anchors Aweigh," "My Country 'Tis of Thee," 'I've Been Working on the Railroad," "Old MacDonald Had a Farm," and "Oh, Susannah."

Generally, this uncalled-for music was described as though it were being re-trieved from a playlist in random fashion. Yet people have also reported that a particular selection could be triggered by some event in their lives. For example, driving along a bridge crossing a river has triggered "Ol' Man River", and walking past a cake in a restaurant has caused "Happy Birthday" to be hallucinated. Colin C. wrote that seeing a door triggered the Rolling Stones's "I see a red door and I want it painted black." And Emily C. remarked that one day, when she was hearing phantom music in an endless stream, it stopped for a few seconds. So she thought, "How wonderful! So that's how silence sounds"—and this thought immediately caused the music to start up again—this time as Simon and Garfunkel's song "The Sound of Silence"!

Although life events may trigger particular musical selections, the hallucinators often insist that they would not voluntarily listen to the music that they "hear," and often dislike the selections that are summoned up by their brains. One correspondent complained that she hallucinated gospel songs that she found offensive, because she was an atheist. Another person who abhorred rap music was plagued by Vanilla Ice's "Ice Ice Baby" for a week. A composer of contemporary art music complained that he hallucinated classical tonal music which he found objectionable, particularly as it interfered with his composing.

Another frequent characteristic of hallucinated music is that a segment gets locked into a continuous loop for minutes, hours—even days or weeks. This was often described as though the music was being played on a scratched or broken record, or on a tape that was constantly being put on rewind. One hallucinator grumbled, "I had to listen to 'Happy Birthday' for about four hours one day." Another complained that she heard "One of These Things Is Not Like the Other" from *Sesame Street* for two straight days. A few people have remarked that the length of the looped segment can change gradually. For example, it might begin with eight measures, which then break off into four measures, then two measures, and then only one measure playing repeatedly.

The continuous looping of hallucinated music, which also occurs when a tune is imagined but not hallucinated, is puzzling. We know that repetition is one of the basic characteristics of music, but why it should occur to such an extreme extent in hallucinations (and also, as we saw earlier, in earworms) is a mystery.

◦ᴖ

The term *hallucination* is derived from the Latin word *alucinari* or *hallucinari*, meaning "to wander in the mind," "to talk idly," or "to rave." The word was used imprecisely for centuries to refer to illusions, delusions, and related experiences. It was not until the work of the French psychiatrist Jean-Étienne Dominique Esquirol in the early nineteenth century that the word came to signify a definite sensation when no external object capable of arousing that sensation was present. In his book *Mental Maladies; A Treatise on Insanity,* published in 1845, Esquirol argued that hallucinations should be clearly distinguished from illusions, which are distortions in the perception of external objects.[2]

The French neurologist Jules Baillarger published the first reports of musical hallucinations in 1846,[3] and the German neurologist Wilhelm Griesinger published further reports in 1867.[4] Yet the medical profession largely ignored these for over a century, partly because people who hallucinated music often kept their experiences to themselves, for fear of being thought crazy. Such fears are generally unfounded; as while hearing speaking voices may be a sign of psychosis (though not necessarily so), musical hallucinations are not generally accompanied by mental illness.

Musical hallucinations are most likely to be experienced by elderly people with hearing loss, though young people with normal hearing may also experience them. Particularly in the younger group, they might reflect an unusual degree of activation in the brain regions that are involved in music perception and memory. On the other hand, young people who are living alone, or who are depressed, are more likely to experience these hallucinations. They can also result from brain injury or disease, such as a stroke, a tumor, or an epileptic focus. A number of prescription medications are thought to produce musical hallucinations—these include tricyclic antidepressants, amphetamines, beta blockers, and carbamazepine—and they can occur following general anesthesia. Excessive amounts of alcohol or recreational drugs can also sometimes give rise to these experiences.

Why should people begin to hallucinate music when their hearing becomes impaired? For people with normal hearing, sounds arriving at the ears are transformed into neural impulses that travel up to the auditory centers of the brain. There, the information is analyzed, transformed, and interpreted, so that the sounds are perceived. The sensory areas of the brain need to have input from the sense organs to function normally. It's believed that when such input is lacking, the brain generates its own images, so that hallucinations result.

This view of musical hallucinations is known as a *release theory*. It has its roots in a theory of brain organization proposed by the nineteenth-century British neurologist Hughlings Jackson.[5] According to Jackson, the brain is constructed of units (groups of neurons) that are hierarchically organized such that higher level units often inhibit lower level ones. When the lower level units don't receive the usual inhibitory signals, their activity is released. Variants of this theory have been suggested to explain hallucinations, though it's assumed that for these experiences, ongoing activity in the sensory areas of the brain is usually inhibited by input from the ears rather than from higher centers. When such input isn't received, this brain activity is released from inhibition, so giving rise to hallucinations. It's also believed that damage to sensory pathways can produce release from inhibition, which is why brain injury and drugs can give rise to hallucinations.

The musical hallucinations that occur among elderly people with impaired hearing have several characteristics in common with the remarkable visual hallucinations that sometimes visit elderly people when they are losing their eyesight. Charles Bonnet, a Swiss naturalist and philosopher, first described these phenomena in 1760 as his grandfather, Charles Lullin, had reported them when his eyesight was failing.[6,7] On one occasion his two granddaughters came to visit, and while they were with him, two young men, magnificently dressed, appeared beside them and later disappeared. On other occasions, Lullin experienced hallucinations of women who were beautifully coiffed, of a swarm of specks that turned into a flight of pigeons, of a rotating wheel that floated through the air, and of objects that grew enormously large, or that were exceedingly small. Other elderly people with failing eyesight have reported seeing beautiful landscapes, large groups of men and women dressed in elaborate clothing, buildings, vehicles, strange-looking animals, and other extraordinary objects.

Visual hallucinations involving persisting or recurring images—termed *hallucinatory palinopsia*—sometimes result from brain damage. For example, one patient, having seen a Santa Claus at a Christmas party, later saw his white beard on the face of everyone around her. Or a patient may see an action sequence, such as someone throwing a ball, and continue to "see" this scene for several minutes— like a looped video. But unlike stuck tunes or musical hallucinations, looped videos are not hallucinated for more than a few minutes, and these phenomena occur very rarely.[8]

⌒

Musical hallucinations can also be due to an abnormally large amount of brain activation, particularly in the temporal lobes. The neurosurgeon Wilder Penfield[9] undertook a monumental study of the experiences of patients that resulted from electrical stimulation of points on the exposed surface of the temporal lobe. The stimulations were performed on patients who were about to undergo surgery for the relief of epilepsy, so as to help localize the origin of their seizures. The patients were fully conscious, and they received only local anesthetics to their scalps, since the brain does not feel pain.

Stimulation of the primary auditory cortex produced impressions of simple sounds such as buzzing or whistling. But stimulation of the neighboring cortex instead produced remarkable dream-like experiences. Patients perceived visual scenes of many descriptions, and heard sounds such as footsteps, a dog barking, or a toilet flushing, and people talking, whispering, shouting, and laughing— and quite frequently they heard music. One patient heard a choir singing "White Christmas"; another heard a group of elderly people singing together; yet another patient heard music from the stage hit *Guys and Dolls* as though it was being played by an orchestra; and yet another heard Mendelssohn's "War March of the Priests" as though it were coming from a radio. These experiences felt real to the patients, even though they were aware that they were in the operating room and could follow events that were happening there.

Penfield noted that similar phantom perceptions, including music, could also occur at the beginning of an epileptic seizure, and he assumed that they were elicited by the same process as was triggered by brain stimulation. One patient reported that before an attack she would always hear a lullaby that her mother had sung: "Hush-a-bye, my baby." Another patient reported that heralding the onset of his attacks he heard an orchestra without vocal accompaniment playing either "I'll Get By" or "You'll Never Know"—songs he had often heard on the radio or at dances. Yet another patient reported that during an attack he heard words set to music that appeared like a radio commercial or a jingle.

Penfield was convinced that in these studies he was not simply activating normal memories, but was instead generating actual reproductions of past experiences. He argued that their vividness and wealth of detail, and the sense of immediacy that accompanied them, distinguished these experiences from the ordinary process of recollection. This conclusion was very similar to those of people who hallucinated music in other situations, and who have had experiences that were far more detailed and striking than those they could have summoned up voluntarily. As one person wrote to me:

> . . . my own perception of what music is playing in my head is that it is a perfect reproduction of all of the song's elements—all the instruments, tone, pitch, rhythm, voices, etc. It is as if the song has left an impression in my memory complete with detailed information about each of its parts . . . a clear facsimile of·a song I have heard before.

It's tempting, therefore, to conclude that music is laid down in memory similarly to downloading a soundtrack onto your cell phone, and that hallucinating a segment of music is like playing a soundtrack from the cell phone. But we know that the process of forming a musical memory must be much more complicated. Suppose you hear a violin playing middle C at a moderate loudness for one second. We can think of a tone as a bundle of attribute values—having a pitch, loudness, duration, timbre, and spatial location. The location is determined by one set of neural circuits, the pitch by another, the timbre by another, and the duration by yet another. The outputs of these different brain circuits then combine so that a synthesized percept results. As described in Chapter 10, experiments on memory for musical tones have shown that separate and distinct modules are involved in storing the different attributes of a tone, and that retrieving a sound from memory involves combining the outputs from the different modules. The same holds true of higher level musical attributes such as intervals, chords, contours, rhythms, tempo, timbre, and so on—these are all analyzed in different neural circuits, and retained in different memory stores. Retrieving music from memory then requires a process of reconstruction in which the values of the different attributes (such as pitch, loudness, timbre, and duration at a low level, and melody, rhythm, and tempo at a higher level) are pulled together. So it seems that both a detailed recording is laid down, and also, in parallel, the music is reconstructed by combining together different attribute values—both when we hear the music, and also when we recall it.

This elaborate structure becomes apparent when we consider the results of brain injury. As discussed in Chapter 10, some patients who have suffered strokes can no longer identify the timbre of a musical instrument such as a piano or a violin, but they recognize melodies and rhythms normally. Other stroke patients are unable to recognize familiar melodies, but have no problem with rhythms or timbres. Yet others have difficulty sustaining a regular beat or recognizing rhythms, but have no difficulty recognizing melodies; and still others report that music sounds "off-key" or "out of tune," though they can recognize the melodies that are being played.

So the evidence from patients with brain damage shows that perception of specific attributes of musical sound can be dissociated by being lost separately from each other. Illusions that occur in normal people can also reflect such dissociations. When two sequences of tones arise simultaneously from different regions of space, these bundles of attribute values may fragment and recombine incorrectly. As described in Chapter 2, this can give rise to striking illusions, such as the scale, glissando, and octave illusions, so that the melodies we hear are quite different from those that are being played.

Musical hallucinations can also reflect such dissociations, so that some aspect of a piece of music may be heard correctly, while other aspects sound altered or corrupted. Hazel Menzies described a vivid example in her email correspondence:

> Weaving in between the long notes in choral-type tones was some snatches of the melody line from Requerdos de l'Alhambra. But it was being played by a chorus of violas at 1 octave below the original guitar melody. Also it was slower. I swear this was not by my own volition. What reason would I choose to do that, after all? I know this guitar piece and have played it in concert in the past; it's one of the most famous pieces in the guitar repertoire and takes no effort on my part to re-play the whole original in my head if I want to "listen" to it—

Distortions of instrument timbre occur quite frequently in hallucinations. Colin C. wrote that in experiencing hallucinations, he heard "occasionally the tone of an instrument I'd never heard before, perhaps some swollen-eggplant sounding combination of an oboe and a bassoon." Others have also described phantom music that sounded as though it was being played by an unknown instrument, or by an instrument that appeared cracked or damaged in some way. And distortions of tempo often occur—usually, when this happens, the speed is too fast, but it can also be too slow, and sometimes it appears to be constantly changing. Other people have complained that the phantom music they hear occasionally appears off-key or out of tune.

At other times, people have hallucinated segments of music that appeared to come from natural instruments, yet would be impossible to play in real life. For example, Hazel Menzies wrote:

Sometimes the whole "chorus" sings one combination of notes in unison for a long time—longer than one real breath would allow.

Jane S. gave another vivid description:

I was wakened out of a sound sleep by hearing a thunderous crash of an enormous chord on a piano that seemed to have been played by a player with maybe 14 fingers. Then—a most incredibly beautiful piano concerto that I could only wish to have been able to remember.

Although these examples might be considered to be musical "mistakes," some hallucinations evidently reflect a sophisticated musical intelligence. Several people have told me that they sometimes hear segments of music in two different genres that are joined together seamlessly so as to make musical sense. Colin C. wrote:

Frequently one song that shares a key with a wildly different song in the same key is bridged together seamlessly. I can, for instance, merge Stevie Wonder and Billy Joel quite well. None of this is controlled, or even entirely wanted. If I wanted to try to intentionally bridge a song with another in my head, I would struggle immensely, if I was even able to complete it.

Other hallucinators have said that they hear phantom music in superimposed layers. The different layers could consist of music in different genres, and yet make perfect sense musically—rather like a successful mashup.[10] These people always wonder at their ability to accomplish such feats of composition, saying that they would have been extremely difficult to achieve by acts of will.

So it seems that the level of brain function at which hallucinations arise is different from the level that normally draws on the characteristics of music in a known musical style to aid in the perceptual construction of musical sound. It's as though in normal function some "executive" at a high level in the auditory system cleans up our perceptual experiences and imaginings in accordance with stored knowledge and expectations, while also inhibiting the creative process. This process of cleaning up is reduced (though not eliminated) at the level where hallucinations are generated.

⌒

While musical hallucinations are rare, hallucinations of speaking voices occur commonly among people who suffer from mental illness. About 70 percent of schizophrenics hear voices that produce words and phrases, and even engage in long conversations. Sometimes the patients hear only one voice, sometimes two, and sometimes many chattering voices. The voices are often critical, abusive, threatening, and use foul language. It's not unusual for them to "command" the patients to perform various acts, which they feel compelled to carry out.

A patient in the psychiatric ward of our local Veterans Administration hospital described to me how he hallucinated two voices, neither of which he could associate with anyone he knew. One was a loud, abusive, male voice that constantly commanded him to throw himself in front of a truck, and so commit suicide in order that he should burn in hell. The other voice appeared less frequently, was soft and female, and said, in counterpoint: "You are not so bad really—you can be good." When I questioned him about the loudness of the voices he "heard," his reply about the male voice was startling: "Suppose you are at an outdoors rock concert standing right next to one of the loudspeakers. Well, it's louder than that." Imagine continuously hearing sounds that are so loud! Even if the content of the "messages" were less threatening, their sheer loudness would be enough to drive you to distraction. Fortunately, hallucinated voices are not generally as loud as this.

The content of the voices heard by people with schizophrenia often reflects delusions. One patient reported that he believed himself to be pursued by unidentified people who spoke to him constantly. As he wrote:

Among these pursuers . . . were evidently were some brothers and sisters, who had inherited from one of their parents, some astounding, unheard of, utterly unbelievable occult powers. Believe-it-or-not, some of them, besides being able to tell a person's thoughts, are also able to project their

magnetic voices—commonly called "radio voices" around here—a distance
of a few miles without talking loud, and without apparent effort, their voices
sounding from that distance as tho heard thru a radio head-set, this being
done without electrical apparatus. This unique, occult power of projecting
their "radio voices" for such long distances, apparently seems to be due to their
natural bodily electricity, of which they have as supernormal amount. Maybe
the iron contained in their red blood corpuscles is magnetised . . . Thus, in
connection with their mind-reading ability, they are able to carry on a conversation with a person over a mile away and out of sight by ascertaining
the person's unspoken thoughts, and then by means of their so-called "radio
voices," answer these thoughts aloud audibly to the person.[11]

People with bipolar disorder and borderline personality disorder sometimes
also hallucinate speaking voices, though less commonly than do schizophrenics.
Posttraumatic stress disorder can bring on phantom speech occasionally as well.
For example, a soldier saw several of his comrades killed when a shell hit his tank,
and a few weeks later he began to "hear" the voices of his dead comrades. The
voices accused him of having betrayed them by leaving and remaining alive, and
they commanded him to join them by committing suicide.[12] Other conditions that
can give rise to hearing voices include epilepsy, excessive consumption of and
withdrawal from alcohol, various medications, traumatic brain injury, strokes, de-
mentia, and brain tumors.

People who are isolated in very stressful situations, such as being imprisoned,
or who have lost their way, sometimes hear phantom voices. In his book *The Third
Man Factor: The Secret to Survival in Extreme Environments*, John Geiger describes
how people, in the midst of traumatic events, and at the limit of human endur-
ance, experienced the close presence of a companion and helper. This "sensed presence" gave them a feeling of protection, and guided them through their ordeals.
The events he recounts were experienced, for example, by 9/11 survivors, polar
explorers, mountaineers, divers, solitary sailors, shipwreck survivors, aviators,
and astronauts. Michael Shermer, in his book *The Believing Brain*, likewise describes many such experiences. In particular, he tells of bizarre hallucinations experienced by cyclists who participated in the 3,000-mile nonstop transcontinental
bicycle Race Across America, in which competitors slept very little and suffered
severe pain and physical exhaustion.[13]

The mountaineer Joe Simpson related such an experience in his book *Touching
the Void*. He described how, separated from his comrade, he endured an agonizing
descent from the mountain Siula Grande with a severely broken leg, taking him
four days to return to base camp. In the latter part of his journey he heard a clear,

sharp voice that encouraged him to keep going—and this probably saved his life.[14] Oliver Sacks, in his book *Hallucinations*, described a similar experience when he was forced to descend a mountain alone with a badly injured leg. He heard a powerful voice that commanded him to continue on his painful journey.[15]

The famous aviator Charles Lindbergh gave a gripping account in his book *The Sprit of St. Louis* of making the first solo nonstop flight between New York and Paris. After 22 hours without sleep, and overcome with fatigue, he had a remarkable experience. As he wrote:

> . . . the fuselage behind me becomes filled with ghostly presences—vaguely outlined forms, transparent, moving, riding weightless with me in the plane . . . These phantoms speak with human voices—friendly, vapor-like shapes, without substance, able to vanish or appear at will . . . First one and then another presses forward to my shoulder to speak above the engine's noise, and then draws back among the group behind. At times, voices come out of the air itself, clear yet far away . . . familiar voices, conversing and advising on my flight, discussing problems of my navigation, reassuring me, giving me messages of importance unattainable in ordinary life.[16]

Historical anecdotes tell a similar story. In his autobiography, the sixteenth-century Florentine sculptor and goldsmith Benvenuto Cellini described his harrowing experiences when he was imprisoned in a dungeon by order of the pope. During his long term of imprisonment he had visions of a "guardian angel" in the form of a beautiful young man who comforted him. Sometimes he could see this "angelic visitor," but sometimes he only heard him speaking in a clear and distinct voice.[17]

Marco Polo, the thirteenth-century explorer whose journey through Asia lasted for 24 years, described experiences that may well have been due to sensory deprivation. He wrote in his book *The Travels of Marco Polo*:[18]

> When a man is riding by night through the desert and something happens to make him loiter and lose touch with his companions, . . . and afterwards he wants to rejoin them, then he hears spirits talking in such a way that they seem to be his companions. Sometimes, indeed, they even call him by name. Often these voices make him stray from the path, so that he never finds it again. And in this way many travellers have been lost and have perished. And sometimes in the night they are conscious of a noise like the clatter of a great cavalcade of riders away from the road; and, believing that they are some of their own company, they go where they hear the noise and, when day breaks,

find that they are victims of an illusion and in an awkward plight. . . . Yes, and even by daylight men hear these spirit voices, and often you fancy you are listening to the strains of many instruments, especially drums, and the clash of arms. For this reason bands of travellers make a point of keeping very close together. (pp. 84–85).

Hallucinations can also arise in situations in which the sense organs are deliberately deprived of the usual amount of stimulation. Beginning with the work of John Lilly in the 1950s, "flotation tanks"—in which the subject floated face up in Epsom-salt-laden water at skin temperature, deprived of sight and sound—were used to study the effects of sensory deprivation. Lacking sensory stimulation, after a while the subjects often began to hallucinate.

Extreme grief can also produce hallucinations. Many people who have recently suffered bereavement have auditory or visual hallucinations of those who have died, and they often hold long "conversations" with them. In one study, people in their early 70s whose spouses had died in the previous year frequently "heard" their spouses' voices, and generally found them comforting.[19]

As a result of its strong association with schizophrenia, the belief that hearing voices is a symptom of insanity is so prevalent that claiming to hear speaking voices puts you at risk of being committed to a mental institution. In an article published in the journal *Science*, titled "On Being Sane in Insane Places," David Rosenhan, a psychology professor at Stanford University, described a remarkable experiment.[20] He and seven other "pseudopatients", with no history of mental illness, presented themselves to the admissions offices of various hospitals with the only complaint that they were hearing voices saying the words "empty," "hollow," and "thud." Although they described no other symptoms, they were all admitted to the psychiatric ward of the hospital at which they had arrived. Once admitted, they reported that they no longer experienced any symptoms, and they behaved quite normally. Nevertheless, the length of their hospitalization before they were discharged ranged from 7 to 52 days!

Despite the popular belief that people who hallucinate speech are probably insane, brief hallucinations of speaking voices or visual apparitions appear more often among healthy people than is generally realized. In the early 1890s, Henry Sidgwick and his colleagues undertook an "International Census of Waking Hallucinations in the Sane," on behalf of the Society for Psychical Research.[21] Most of the respondents were English, but others were from the United States, Russia, or Brazil. The researchers were careful to exclude people who had an obvious physical or mental illness, and also to exclude hallucinations that were experienced on the threshold of sleep—since such occurrences are not unusual. Nevertheless, 2.9 percent of the respondents reported that they had experienced hearing voices. A later study concluded that roughly 1.5 percent of people with no impairment of function or distress reported that they heard voices.[22]

Cultural factors, in particular an emphasis on religion, play an important role in voice hearing. In the Old Testament it is stated that God conversed with Adam in the Garden of Eden. He also spoke to Moses from the burning bush, and dictated the Ten Commandments to him. Abraham, Isaiah, Jeremiah, Ezekiel, Job, and Elijah all experienced divine voices. In The New Testament it is stated that Jesus conversed with the Devil and refused to be tempted, and that St. Paul was converted to Christianity by a voice that spoke to him.[23]

In accordance with with religious tradition, many devout spiritual leaders have heard voices. In the thirteenth century, St. Francis of Assisi claimed to have received his "calling" to the religious life from the voice of God. In the fifteenth century, Joan of Arc asserted that she heard the voices of saints commanding her to help the king of France win back his kingdom from the English invaders. Other religious figures who heard voices included St. Augustine, Hildegaard of Bingen, St. Thomas Aquinas, St. Catherine of Siena, and St. Teresa of Avila. In the present day, evangelical Christians who engage extensively in prayer and other spiritual exercises sometimes see visions and hear voices.[24]

When people hallucinate speaking voices, they do not generally hallucinate music as part of their experience. And although people who hallucinate music may sometimes hear speech-like sounds, these generally appear to be mumbled and indistinct, and the words can be difficult to make out. So it appears that the brain circuits that generate musical and verbal hallucinations are largely distinct and separate.

There are rare accounts of composers who hallucinated music, and who even composed music that was inspired by hallucinations. Robert Schumann suffered intermittently from "unusual auditory disturbances," which became severe toward the end of his life. He also became so absorbed in his internal music that when he was conducting, the orchestra members had difficulty following him. At this point he clearly had difficulty distinguishing illusion from reality, His fellow composer Joseph Joachim described how Schumann once expressed disappointment at being unable to hear a "magical horn solo" during a performance—yet the horn player hadn't even arrived!

Late in his life Schumann was hounded by auditory disturbances that initially took the form of continuous pitches, and then developed into entire compositions that appeared "in splendid harmonizations" performed by a "distant wind-band." In February 1854 he rose in the middle of the night to jot down a theme that was "dictated by the angels" and that he ascribed to the spirit of Schubert. Yet by the next morning the angelic voices had turned into the voices of demons, who sang "hideous music" and threatened to "cast him into hell." Good and evil spirits hounded him during the following week, and at this time he wrote his last composition, a set of five variations on the angelic theme, which are known as *Geistervariationen* ("Ghost Variations"). Soon after this he was admitted to an insane asylum, and he died there a few years later.[25]

The composer Bedřich Smetana also developed musical hallucinations toward the end of his life—these were most likely the result of syphilis, and heralded the onset of deafness. He described one hallucination as follows:

> as I was walking in the early evening hours through the woods . . . I suddenly heard such moving and ingenious notes being lured from a flute that I stood still and looked around me, trying to see where such an excellent flute player was hiding. Nowhere, however, could I see a living soul. I passed this over without noticing; when this happened again next day, I kept to my room, but the illusion repeated itself later in a closed room and so I went to seek advice from the doctor.[26]

The great pianist Sviatoslav Richter suffered from hallucinations. In his autobiography he wrote of a period in 1974:

> This was accompanied by a type of auditory hallucination that tortured me for months on end, day and night, even while I was asleep. I started to hear a recurrent musical phrase a few bars long, violently rhythmical and rising in pitch. It was based on a chord of a diminished seventh. In the cold light of day I tried to work out what it meant, even though the torment was permanent, even telling myself that such a phenomenon might be of interest to medical science. But try telling doctors about chords of a diminished seventh! Sometimes I would lie awake all night trying to work out what I was hearing—I wasn't hearing it, but just *thought* I was—or to work out its pitch. I was forever trying to identify the notes and these primitive harmonies and to correct them, as it was the most frightful nonsense—ta raaa ra riii ri rii— going through my head in every conceivable key.[27]

The compositions of Charles Ives are astonishingly reminiscent of hallucinated music. Steven Budiansky, in his remarkable book, *Mad Music: Charles Ives, the Nostalgic Rebel*,[28] describes some of the characteristics of Ives's music, such as distortions of memory, obsessive recollections of familiar tunes, inclusion of environmental sounds, wrong notes, wrong timing, fragments that are off key or out of tune—all of which occur in hallucinations. Ives's composition *Putnam's Camp* is a particularly good example. It includes patriotic and religious tunes, loud, intrusive music juxtaposed and overlaid with fragments in entirely different styles, inappropriate dissonances, strange mixtures of timbres, and segments that wander off key and out of tune. Ives didn't say that he experienced musical hallucinations; however, he was very reticent about his personal life, and if he did hallucinate music, he would have been likely to keep this to himself.

Other expert musicians are skeptical of this interpretation of Ives's compositions. The composer Michael A. Levine and the musicologist Peter Burkholder have both argued that although Ives's compositions are very similar to hallucinated music, they more likely reflect the composer's extraordinary ability to recapture in detail the experiences he had when listening to musical pieces in real life.[29] Ives appeared to be able to register all the sounds that reached him at the same time, rather than listening selectively to music or to a conversation, and he could also remember in remarkable detail the mixtures of sounds that he heard. He remarked that his hearing process was different from that of other people, having once declared "Are my ears on wrong?"[30] In addition, as a child, his father had encouraged him to experiment with composing music including notes that were out of tune, and segments that were out of key, or that were simultaneously in unrelated keys.

We shall never know whether Ives hallucinated music, of course, and some composers have auditory memories that are sufficiently vivid that they approach hallucinations. Indeed, the borderline between musical hallucinations and strong imagery for music is sometimes blurred, particularly for some expert musicians. A distinguished composer wrote to me:

> I almost never confuse real and imagined music with one another, but there are exceptions. A few weeks ago, I stopped in a corridor at school to listen more closely to a passage in a piece I had recently completed that was running through my mind. At least I thought it was running through my mind— it was a full minute before I realized I was hearing musicians who were rehearsing the piece in a room nearby.

Taken more generally, musical hallucinations have several characteristics in common with earworms, so that, to some extent, the two are likely to have the same neurological underpinnings. For example, they both tend to appear out of the blue, and yet can be triggered by some event in the environment. Also, for many people, the hallucinations are snippets of well-known tunes that repeat continuously—and this is also characteristic of earworms.

On the other hand, there are important differences between these two types of internal music, in that the final conscious perception is radically different in the two cases—one is clearly experienced as a thought and the other is phenomenally indistinguishable from real sound. People don't experience earworms as having loudness—whereas people who experience musical hallucinations are quick to describe their loudness in detail, particularly if they are very loud and outside their conscious control. Also, while most earworms are generally impoverished—frequently they are only repeating segments of well-known melodies—several people who experience musical hallucinations hear them in elaborate detail, including features such as melody, harmonic structure, timbre, and tempo, and they often describe them as out of tune or as being played by an unknown instrument. Researchers haven't yet identified the neurological mechanisms that distinguish these two kinds of internal music.

❧

So far we have been considering hallucinations of music and speech. In the following chapter we consider an illusion—the "speech-to-song illusion"—in which a repeated spoken phrase is perceptually transformed so as to be heard as sung rather than spoken. Beginning with this illusion, we discuss how speech and song are related, how they differ, and how this strange phenomenon might be explained.

# 10

## The Speech-to-Song Illusion

### CROSSING THE BORDERLINE BETWEEN SPEECH AND SONG

... whatever speech I hear, no matter who is speaking, ... my brain immediately sets to working out a musical exposition for this speech.[1]

—MODEST MUSSORGSKY, 1868

ONE AFTERNOON IN the summer of 1995, a curious incident occurred. I was putting the finishing touches on a CD that I was preparing on music and the brain, and was fine-tuning my spoken commentary. In order to detect and correct small glitches in the recorded speech, I was looping phrases so that I could hear them over and over again. The opening commentary contains the following sentence: "The sounds as they appear to you are not only different from those that are really present, but they sometimes behave so strangely as to seem quite impossible." I put one of the phrases in the sentence— *sometimes behave so strangely*—on a loop, began working on something else, and forgot about it. Suddenly it appeared to me that a strange woman had entered the room and was *singing*! After glancing around and finding no one there, I realized that I was hearing *my own voice* repetitively producing this phrase—but now, instead of hearing speech, it appeared to me that a sung melody was spilling out of the loudspeaker—as in Figure 10.1. So the phrase had perceptually morphed from speech into song by the simple process of repetition.

sometimes    behave    so    strangely

FIGURE 10.1. The spoken phrase "sometimes behave so strangely" as it appears to be sung after it has been repeated several times. From Deutsch (2003).

The *speech-to-song illusion* (as I named it) is indeed bizarre. It occurs without altering the signal in any way, and without any context provided by other sounds,

**Speech-to-Song Illusion**

http://dianadeutsch.net/ch10ex01

but simply through repeating the same phrase several times over. As a further twist, when the full sentence is again played, it starts out by sounding exactly like normal speech, just as before. But when it comes to the phrase that had been repeated—*sometimes behave so strangely*—it appears suddenly to burst into song. And this transformation is amazingly long-lasting; once you've heard the phrase repeatedly so that it sounds like song, it continues to sound like song even after months or years have elapsed.

You can hear this strange illusion in the "Speech-to-Song" module. It has no obvious explanation in terms of current scientific thinking about the neural underpinnings of speech and music. It's generally assumed that a phrase is heard as either sung or spoken depending on its physical characteristics. Speech consists of pitch glides that are often steep, and of rapid changes in loudness and timbre. In contrast, song consists largely of well-defined musical notes, and so of more stable pitches, and these combine to form melodies and rhythms.

Given the physical differences that generally exist between speech and song, scientists have focused on their physical properties in attempting to understand how they are perceived. Around the middle of the twentieth century the view emerged that when we listen to speech, the signal bypasses the usual neural pathways for processing sound, and is analyzed in an independent module that is located in the left hemisphere of the brain and that excludes from analysis other sounds such as music. This view was soon joined by the further view that music is also the function of an independent module, in this case located in the right hemisphere, and that excludes speech sounds from analysis.[2,3]

Yet the claim that music and speech should be considered distinct and separate is difficult to reconcile with the many types of vocalization that fall at the boundary between speech and song. These include religious chants, incantations, opera recitative, *Sprechstimme*, tone languages, whistled languages, and rap

music—all indicating that speech and music are intertwined. The claim also con-
flicts with the view taken by philosophers and musicians throughout the ages that
strong linkages exist between speech and music.

In the nineteenth century, the British philosopher Herbert Spencer argued that
a continuum extends from ordinary speech at one end to song at the other, with
emotional and heavily intoned speech in between. In an article entitled "On the
Origin and Function of Music,"[4] Spencer pointed out that in speech, excitement is
conveyed by rapid pitch variations, depressed emotions by slowness of articulation,
anger and joy by high pitch ranges, and surprise by contrasting pitches. He wrote:

> what we regard as the distinctive traits of song, are simply the traits of emo-
> tional speech intensified and systematized. In respect of its general character-
> istics, we think it has been made clear that vocal music, and by consequence
> all music, is an idealization of the natural language of passion . . . vocal music
> originally diverged from emotional speech in a gradual, unobtrusive manner.

Spencer also argued that pitch variations in normal speech consist of narrow
intervals, but when people become emotional, the intervals in their speech
become wider.

> While calm speech is comparatively monotonous, emotion makes use of
> fifths, octaves, and even wider intervals. Listen to any one narrating or re-
> peating something in which he has no interest, and his voice will not wander
> more than two or three notes above or below his medium note, and that by
> small steps; but when he comes to some exciting event he will be heard not
> only to use the higher and lower notes of his register, but to go from one to
> the other by larger leaps.

For hundreds of years, composers have also believed that expressivity in music
can be derived from inflections in speech, and they have incorporated into their
compositions various characteristics of emotional expression in spoken utterances. Notable among these were the Renaissance composers Carlo Gesualdo and
Claudio Monteverdi.

In the nineteenth century, the Russian composer Modest Mussorgsky became
convinced that speech and song fall at the opposite ends of a continuum that was
derived from speech intonation. In his compositions, he began drawing on conversations that he overheard, employing the intervals, timing, and loudness variations of natural speech, and amplifying intonation curves into melodic lines. As
he wrote to his friend Nikolai Rimsky-Korsakov:

whatever speech I hear, no matter who is speaking . . . my brain immediately
sets to working out a musical exposition for this speech.[5]

And later he wrote:

My music must be an artistic reproduction of human speech in all its subtlest
inflections, that is, the *sounds of human speech* as the external manifestation
of thought and feeling must, without exaggeration and forcing, become
truthful, precise *music*.[6]

A particularly fine example of Mussorgsky's compositional style involving
speech is his song cycle *The Nursery*. This consists of a collage of "scenes," that
Mussorgsky composed to his own lyrics, and that enable us to listen in to the im-
aginative world experienced by children—a child is frightened by a fairy tale, startled by a huge beetle, rides a hobby horse, then falls off the horse, and is comforted
before taking off again.

The Czech composer Leoš Janáček was also preoccupied with the idea of music
as heavily intoned speech. As he wrote, "A good preparatory study for opera composers is careful eavesdropping of folk speech melodies."[7] He was himself a sea-
soned eavesdropper, and spent more than three decades transcribing thousands
of speech segments into standard musical notation. For example, he notated a
brief conversation between a teacher and his student, the sound of an old woman
complaining to her butcher, the taunt of a child, and the calls of street vendors.

At the turn of the twentieth century, other composers experimented with a
technique of vocal production known as *Sprechstimme*, which combines elements
of both song and speech. Although *Sprechstimme* has been used in different ways,
often the rhythms of speech are retained, but the pitches are only approximated,
and sung notes often consist of pitch glides rather than steady pitches. The opera
*Königskinder*, by Engelbert Humperdinck, contains passages that are somewhere between singing and speech. Also at around that time, Arnold Schoenberg composed
*Pierrot Lunaire* to a cycle of poems by Albert Giraud. Further works featuring blends
of speech and music include *L'Histoire du soldat*—with the music composed by Igor
Stravinsky, and the libretto written by C. F. Ramuz. A series of verses by Edith Sitwell
titled *Façade—An Entertainment* represents a particularly engaging use of this technique. Her declaimed verses are accompanied by music by William Walton.

Later composers incorporated actual speech samples into their pieces. In
*Different Trains*, Steve Reich combined fragments of recorded speech together with
a string quartet that reproduced the melodic and rhythmic contours of the speech
fragments. The musical quality of the speech is emphasized by instrumental

doublings of the melodic segments, as well as by instruments playing these segments separately. In earlier work by Reich, such as *Come Out*, a looped fragment of speech is initially played on two channels in unison. The voices move increasingly out of synchrony, then split into four channels, and then into eight; with the result that ultimately rhythmic and tonal patterns emerge strongly.

In the piece *John Somebody*, by the composer-guitarist Scott Johnson, short fragments of a basic utterance are looped in parallel channels. The speech is joined by instruments, which at times double the spoken fragments with the same melodic material, and at other times play this material separately. As in *Different Trains*, the use of musical instruments to double pitches in the speech enhances the musical salience of the spoken words.

Beginning in the 1980s, when sampling technology became widely available, music that incorporated the extensive use of spoken phrases became very popular. This is particularly true of rap or hip-hop music—a mass-market form of music that includes the chanting of rhythmic and rhyming speech.

⟶

Despite the strong bonds between speech and music that are reflected in much performed music, around the mid-twentieth century researchers began to espouse the view that these two forms of communication were the product of separate brain systems that each had a unique and dedicated neural architecture These two systems (or "modules") were held to be encapsulated from each other and from the rest of the auditory system. It was further claimed that in most right-handers the mechanism underlying speech was situated in the left (dominant) hemisphere, while the mechanism underlying music was situated in the right (non-dominant) hemisphere.

Proponents of the modular view of speech processing referred to the early studies of Broca in the nineteenth century (described in Chapter 1), and to later clinical studies that confirmed his findings. This work showed that certain patients with lesions in the left hemisphere had severe speech impairments, whereas those with comparable lesions in the right hemisphere spoke without difficulty. The researchers also argued that the ways in which we process speech sounds differ in several respects from those in which we process other sounds in the environment.

The modular view of musical processing received strong support from a remarkable case study entitled "Aphasia in a Composer,"[8] published by Alexander Luria and his colleagues at Moscow University. The subject of this study was Vissarion Shebalin—a distinguished composer who had been a director of the Moscow Conservatory. At age 57, Shebalin suffered a massive stroke that produced extensive damage to the temporal and parietal regions of his left hemisphere, leaving his right hemisphere unimpaired. The stroke left Shebalin profoundly aphasic, so that his ability to speak and understand the speech of others became seriously compromised. Three years following his stroke, the composer tried haltingly to explain his difficulties by saying: "The words . . . do I really hear them? But I am sure . . . not so clear . . . I can't grasp them . . . Sometimes—yes . . . But I can't grasp the meaning. I don't know what it is."

Yet despite his profound aphasia, Shebalin created many new compositions, which other distinguished musicians considered to be as high in quality as those he had created earlier. The composer Dmitri Shostakovich wrote:

> Shebalin's Fifth Symphony is a brilliant creative work, filled with the highest emotions, optimistic and full of life. The symphony composed during his illness is a creation of a great master.

Because Shebalin had suffered massive damage to his left hemisphere and not his right, Luria concluded that the processing of speech and music must take place in different neural structures.

Another remarkable case of severe aphasia without amusia involved a distinguished musician, who for many years conducted the orchestra at La Scala in Milan. During a concert tour, he suffered a massive stroke that affected his left hemisphere, including the temporal lobe. The stroke left him severely aphasic, so that two years later his speech was confined to brief expressions such as "It's difficult" and "I know it." He also had serious trouble understanding speech, was unable to write, and for the most part couldn't make appropriate gestures. Yet his musical capacities remained largely intact, and most remarkably, he continued to perform brilliantly as a conductor, with one music critic writing:

> The 71-year-old Venetian conductor has shown that his extraordinary gifts remain untarnished, gifts of exceptional musical sense, of easy gesture perfectly suited to the music of today, and of warm, communicative humanity.[9]

In contrast, strokes in other patients have produced impairments in various specific aspects of musical function, leaving their speech intact. One amateur musician, who was studied at the University of Parma, suffered a mild stroke that damaged his right temporal lobe. His understanding of speech was undiminished, but he had difficulty identifying the sounds of musical instruments, and even of his own voice. Yet he could recognize other aspects of music such as melodies, rhythms, and tempi, and could sing familiar tunes; so his deficit appeared to be confined largely to the perception of timbre.[10]

A further contrast was provided by a left-handed amateur musician who had suffered damage to his left temporal lobe that produced impairments in recognizing tonal structures, while his processing of speech and environmental sounds remained normal. He complained, "I can't hear any musicality: All the notes sound the same" and "Singing sounds like shouting to me." He could compare rhythms correctly and recognize the timbre of musical instruments, as well as other sounds such as coughing, laughing, and animal noises; he could even distinguish foreign accents. But he had serious difficulty in recognizing songs, even those with which he was very familiar, such as the national anthem, and "'O sole mio."[11]

Even more remarkable are people with lifelong difficulties in perceiving and comprehending music, yet who are in other ways normal or even brilliant. The inability to recognize simple tunes has been described using a variety of terms, including *tone-deafness, amelodia,* and *congenital amusia.* Charles Darwin described his own inability to make sense of music in his autobiography. He wrote of his years as a student in Cambridge:

> I acquired a strong taste for music, and used very often to time my walks so as to hear on week days the anthem at King's College Chapel. This gave me intense pleasure, so that my backbone would sometimes shiver. I am sure that there was no affectation or mere imitation in this taste, for I used generally to go by myself to King's College, and I sometimes hired the chorister boys to sing in my rooms. Nevertheless, I am so utterly destitute of an ear, that I cannot perceive a discord, or keep time and hum a tune correctly; and it is a mystery to me how I could possibly have derived pleasure from music.
>
> My musical friends soon perceived my state, and sometimes amused themselves by making me pass an examination, which consisted in ascertaining how many tunes I could recognize, when they were played rather more quickly or slowly than usual. "God save the King," when thus played, was a sore puzzle. There was another man with almost as bad an ear as I had, and strange to say he played a little on the flute. Once I had the triumph of beating him in one of our musical examinations.[12]

In the late nineteenth century, the distinguished naturalist and author Charles Grant Allen published an article in the journal *Mind,* entitled "Note-Deafness."[13] He pointed out that just as some people suffer from color blindness, others have an analogous condition in hearing. As he noted:

> Not a few men and women are incapable of distinguishing in consciousness between the sounds of any two tones lying within the compass of about half an octave (or even more) from one another.

In particular, Allen described an educated young man, of whom he wrote:

> If any two adjacent notes upon the piano be struck, he is quite incapable of perceiving any difference between them. After careful and deliberate comparison, many times repeated, he believes the two sounds to be exactly alike. . . . Further, if any note, say C is played on the piano, and another note at a considerable interval, say E or A in the same octave, is subsequently played, he cannot notice any difference between them. As the interval enlarges to an octave or more, as from C to C′ or A′, he becomes gradually aware of a difference in pitch.

Grant Allen also pointed out that while his subject had severe difficulties in processing pitch, he was acutely sensitive to the timbres of various musical instruments:

> A piano is, for him, a musical tone, plus a thud and a sound of wire-works; a fiddle is a musical tone plus a scraping of resin and cat-gut; while an organ is a musical tone, plus a puff of air and an indistinct noise of bellows.

The prevalence of amusia is estimated to be roughly 4 percent in the general population (at least of nontone language speakers), which is comparable overall to the prevalence of color blindness.[14] A detailed study concerned an intelligent woman in her early forties, whom the authors referred to as Monica. Her hearing was normal, and she had taken music lessons in childhood. Yet she was quite unable to discriminate between melodies based on pitch contour or interval, and she had severe trouble discriminating pitch differences between pairs of tones. Monica also couldn't identify well-known melodies that other people recognized easily. Yet she had no difficulty in identifying spoken lyrics, voices, or environmental sounds.[15]

Neuroanatomical studies have shown that tone deaf people have reduced connectivity in the arcuate fasciculus—a highway of white matter that connects the temporal and frontal regions of the brain. Right-handers with disrupted connectivity in the arcuate fasciculus of the left hemisphere are unable to repeat back words and phrases they hear—a condition known as *conduction aphasia*. Psyche Loui and her colleagues at Harvard Medical School found that tone deaf people also have reduced connectivity in the arcuate fasciculus, in this case particularly in the right hemisphere.[16]

In stark contrast to people who have lifelong musical disabilities but are otherwise normal, a few people have substantial impairments in general function, yet are remarkably fine musicians. For example, Rex Lewis-Clack was born blind with a condition known as septo-optic dysplasia, and is severely brain damaged—however, his musical abilities are extraordinary. When Rex was eight years old, CBS Correspondent Lesley Stahl interviewed him for the program *60 Minutes*. He couldn't carry out a conversation or dress himself, yet when Stahl played him a fairly complicated piece on the piano—which he had never heard before—he played it back without difficulty. Stahl interviewed Rex again when he was thirteen years old. His music teacher sang and played him a long and sophisticated song that he didn't know (Schubert's "Ave Maria") and he immediately played and sang it back. At that time he was composing and improvising music, and could easily transpose pieces into different keys, and transform them into different musical styles. Stahl described Rex as "a study in contrasts . . . a mysterious combination of blindness, mental disability, and musical genius."[17]

As earlier described, given the dissociated impairments between music and speech processing, researchers have surmised that these two forms of communication are governed by separate and independent modules, each of which is localized in a specific region of the brain. So when sophisticated brain mapping techniques, such as fMRI and MEG,[18] were developed in the late twentieth century, the hope arose that maps of brain regions underlying speech and music could be created by recording patterns of neural activation in normal subjects. It was further hoped that these maps would be as expected from impairments in speech and music resulting from brain damage: When listening to spoken phrases, certain "speech areas" in the left hemisphere would light up, and when listening to music, certain "music areas" in the right hemisphere would light up instead.

However, brain imaging experiments didn't initially produce the hoped-for results. Researchers such as Stefan Koelsch[19] and Aniruddh Patel[20] found that listening to music produced activity in many parts of the brain, including Broca's and Wernicke's areas in the left hemisphere, which had been thought to be particularly involved in speech. Koelsch and his colleagues boldly entitled an article which they published in 2002, "Bach Speaks: A Cortical Language-Network Serves the Processing of Music." Other fMRI studies have demonstrated the involvement of both speech and music in the frontal, temporal, and parietal lobes of both hemispheres, and in the basal ganglia, thalamus, and cerebellum, leading to the conclusion that they involve vastly distributed brain networks (Figure 10.2).

On the other hand, there is a problem with interpreting the results from standard fMRI studies. The units that are used for this type of analysis, called "voxels", have a coarse spatial resolution, and reflect the responses of hundreds of thousands—or even millions—of neurons. So if different populations of neurons overlap in a brain region, they should be difficult to isolate. The neuroscientists

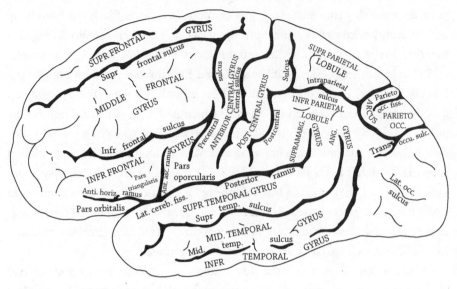

FIGURE 10.2. Different regions of the cerebral cortex.

Sam Norman-Haignere, Nancy Kanwisher, and Josh McDermott at MIT have devised a mathematical technique for isolating overlapping neural populations from fMRI responses. Using their new method of analysis, they revealed distinct neural pathways for music and speech in neural populations that lie beyond the primary auditory cortex.[21] So overall, the brain imaging studies, taken together with studies of brain-damaged patients, lead to the conclusion that some neural circuits specifically subserve speech or music, while other circuits subserve both forms of communication.

But the question of modularity needs to be taken further: Neither speech nor music should be considered a monolithic whole; rather, we can assume that certain circuits are unique to specific aspects of speech (such as semantics or syntax); that other circuits are unique to specific aspects of music (such as melody or rhythm); and that still other circuits are involved in both these forms of communication.

From another perspective, we must conclude that in perceiving either music or speech, we draw on a set of modules, each subserving a particular attribute, and that we link together the outputs of these modules so that a combined percept emerges. The evidence from patients with brain damage is important here. Presumably the damage suffered by these patients compromised the functioning of one or more modules that were involved in processing speech or music, leaving other speech- or music-related modules intact.

As we saw in Chapter 9, further evidence for this modular view comes from people who hallucinate music. Often the music they "hear" can appear accurate in fine detail, except that one attribute may be perceived incorrectly or sound distorted. For example, the hallucinated music may be different from the original in pitch range, loudness, tempo, or it may appear as though played by an unfamiliar instrument.

Short-term memory for tones appears to be subserved by several specialized modules. In one study I found that the ability to tell whether two test tones were identical in pitch was strongly disrupted by an intervening sequence of other tones, but not by a sequence of spoken words.[22] This is demonstrated in the "Tones and Words in Memory" module. From this experiment, I concluded that memory for pitch is retained in a specialized system that disregards other aspects of sound.[23] Catherine Semal and Laurent Demany,[24] using a similar paradigm, found that an intervening sequence of tones whose timbres differed from those of the test tones didn't interfere with recognition of the test tone pitches. Other studies have obtained analogous findings with respect to short-term memory for timbre.[25] So this strengthens the view that speech and music are each subserved by a set of modules, and that some (though not all) of the modules are dedicated to processing a particular attribute of sound.

**Tones and Words in Memory**

http://dianadeutsch.net/ch10ex02

The view of the auditory system as a set of modules, each with a specialized function, with the whole being influenced by general factors, is similar to that expressed by Steven Pinker, in his book *How the Mind Works*. Pinker wrote:

> The mind has to be built out of specialized parts because it has to solve specialized problems. Only an angel could be a general problem solver; we mortals have to make fallible guesses from fragmentary information. Each of our mental modules solves its unsolvable problem by a leap of faith about how the world works, by making assumptions that are indispensable but indefensible—the only defense being that the assumptions worked well enough in the world of our ancestors. (p. 30)[26]

The organization of information in terms of a hierarchy of modules confers a significant advantage. If a problem occurs in a complex system that has a flat or non-hierarchical structure, and part of this structure is damaged, there is a substantial possibility that the entire system will be compromised. On the other hand, when a system is composed of a set of independent modules, if one module is damaged the others can still remain intact. The computer scientist Herbert Simon illustrated this point strikingly with a parable of two watchmakers who use different algorithms to create their devices—one flat and non-hierarchical and the other modular. If the "non-hierarchical watchmaker" makes an error, the damage could be very substantial, since the different parts of the entire system are linked together. However, the "modular watchmaker" is able to repair the damaged part of the system, without compromising the remaining parts. Another advantage of a modular system is that a module can be "tinkered with," and so be improved without necessarily disturbing the other components of the system. This provides an evolutionary advantage.[27]

The evolutionary biologist Tecumseh Fitch has argued for an approach with respect to language that is similar to the one taken here. As he wrote:

> Rather than viewing language as a monolithic whole, I treat it as a complex system made up of several independent subsystems each of which has a different function and may have a different neural and genetic substrate, and, potentially, a different evolutionary history from the others. (p. 5).[28]

⌒

The speech-to-song illusion led me to consider the puzzle of how language and music can involve different parts of the brain, and yet be intimately related. As a first step toward unraveling this mystery, I carried out a study with my colleagues Trevor Henthorn and Rachael Lapidis in which we played subjects the spoken phrase "sometimes behave so strangely" (as recorded on my CD *Phantom Words, and Other Curiosities*) under different conditions.[29]

We first tested three groups of subjects, with each group receiving a different condition. The subjects all heard the full sentence as it occurs on the CD, followed by ten presentations of the phrase "sometimes behave so strangely" that was embedded in it. Following each presentation, the subjects judged how the phrase had sounded on a five-point scale ranging from *exactly like speech* (1) to *exactly like song* (5). We then analyzed the subjects' ratings of the first and last presentations of the phrase.

In all conditions, the first and last presentations were identical to the original one. In the first condition, the intervening phrases were also identical to the original. In the second condition, they were transposed slightly (2/3 semitone or 1 1/3 semitone up or down), so that they consisted of different pitches, but the pitch relationships were preserved. In the third condition, the intervening phrases were not transposed, but the syllables were presented in jumbled orderings.

As shown in Figure 10.3, when the original and intervening phrases were identical, the subjects' judgments of the final phrase moved solidly from speech to

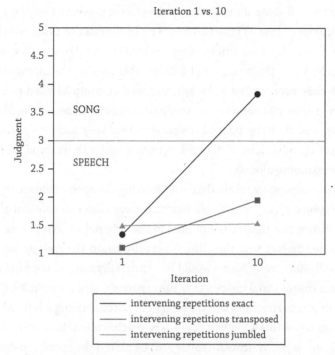

Iteration 1 vs. 10

FIGURE 10.3. Listeners' judgments of the spoken phrase "sometimes behave so strangely" following ten repetitions of the phrase. When the repetitions were exact, the phrase was heard solidly as song. When the phrase was transposed slightly on each repetition, the phrase continued to be heard as speech. And when the syllables were presented in jumbled orderings, the phrase also continued to be heard as speech. Based on data from Deutsch, Henthorn, & Lapidis (2011).

song. When the repeating phrases were transposed, perception of the final phrase moved slightly toward song, but remained solidly in the speech region. When the syllables in the repeating phrases were jumbled, the final phrase was judged solidly to be speech. So for the perceptual transformation from speech to song to occur, the spoken phrase needed to be repeated exactly, without transposition, and without changing the order of the syllables.

We next recruited eleven female subjects who had sung in choirs or choruses, and we tested them individually. We asked them to listen to the full sentence, followed by the phrase repeated ten times, and then to reproduce the phrase as they finally heard it. All these subjects reproduced the final phrase as song, with the melody as shown in Figure 10.1. We then recruited another group of eleven subjects on the same basis, and asked them

**Spoken Phrase Reproduced after Repeated Presentation**

http://dianadeutsch.net/ch10ex03

to listen to the full sentence followed by the phrase presented *only once*, and then to reproduce the phrase as they had heard it. In contrast to those who had heard the phrase repeatedly, these subjects reproduced the final phrase as speech, using a substantially lower pitch range and considerably varying the timing and contour with which they pronounced it. Finally, the second group listened to a recording of the phrase that was sung by my study co-author Rachael as she had heard it many times, and they reproduced the sung phrase very accurately. You can hear the subjects' reproductions of the spoken phrase under these different conditions in the accompanying module.

In further analyses, we found that after hearing the spoken phrase repeated ten times, the subjects sang it with pitches that were closer to those of the melody shown on Figure 10.1 than to the pitches in the original spoken phrase. This happened despite the fact that they hadn't actually heard the melody. So in experiencing this illusion, the subjects would have mentally warped the pitches so as to conform to a simple tonal melody that they created in their own minds![30]

Remarkably, this transformation doesn't only occur among adults with musical training. I discussed this with Walt Boyer, who teaches music in Atwater Elementary School in Shorewood, Wisconsin. Intrigued, he played the speech-to-song illusion to his class of fifth graders (10- and 11-year-olds) without first telling them what they might hear, and he videotaped their responses. As I had done, he began with the full sentence, and then played the spoken phrase repeatedly. In the middle of the repetitions the kids became intrigued, so he said "Try it—go ahead!" whereupon they sang along with the phrase—first tentatively, and then with gusto—and in tune! Mr. Boyer recorded the event on a video that is embedded in the accompanying module. You can see that as the repetition continues, the kids perk up and laugh, and some even make dancing movements.

**Children Responding to the Speech-to-Song Illusion**

http://dianadeutsch.net/ch10ex04

So what cues do we use to determine whether a phrase is being sung or spoken? In general, there are many physical characteristics that distinguish speech from song. For example, song tends to consist largely of flat tones, and so of relatively stable pitches, whereas speech often consists of pitch glides. And vowels in song tend to be of longer duration than those in speech. But in the speech-to-song illusion, the physical characteristics of the phrase don't change—the phrase is repeated exactly. So here, repetition must here be acting as a strong cue that the phrase is being sung rather than spoken. And indeed, repetition is a very powerful feature of music, and it appears in all known musical cultures.

Everyday speech provides a striking contrast. In normal conversation, a spoken phrase that's repeated several times in succession appears quite bizarre, and would likely cause us to wonder whether the speaker is joking, or being sarcastic. The incongruity of repetition in conversational speech is illustrated in an email that the composer Michael A. Levine sent me concerning my CDs. He wrote:

> I started listening to them and they usually playback fine. But they sometimes behave so strangely. Sometimes behave so strangely. Sometimes behave so strangely. Sometimes behave so strangely. Sometimes behave so strangely. Sometimes behave so strangely. Sometimes behave so strangely . . . . . . [31]

It is easy to understand why there is a difference in the amount of repetition between speech and song. The main purpose of conversational speech is to transmit information about the world, and spoken words are useful in that they represent objects and events—the sounds themselves are not informative. Indeed, conversational speech that contains unnecessary repetition is counterproductive, since it slows down the transmission of information.

Given the basic purpose of speech, when we recall a conversation, we're remarkably poor at remembering the precise sequence of words that were spoken—instead we remember the gist, or general meaning, of what was said. In contrast, when we hear a piece of music, it makes no sense to attempt to paraphrase or summarize it, since it carries no semantic meaning—the sounds and sound patterns stand for themselves. So it's understandable that music, but not conversational speech, should contain a substantial amount of repetition.

⌂

While repetition of phrases occurs only infrequently in daily conversations, it is used to striking effect in other forms of speech, such as oratory or poetry. In Shakespeare's *Julius Caesar*, when Brutus addresses the crowd to justify Caesar's assassination, he employs repetition as a device to convince the crowd:

> Who is here so base that would be a bondman? *If any, speak; for him have I offended.* Who is here so rude that would not be a Roman? *If any, speak; for him have I offended.* Who is here so vile that will not love his country? *If any, speak; for him have I offended.* I pause for a reply.[32]

In the twentieth century, Winston Churchill was famous for his powers of oratory. In particular, as prime minister of Great Britain during World War II, he used repetition to inspire feelings of hope and patriotism in his speech to the House of Commons on June 4, 1940, following the evacuation of Dunkirk:

*We shall fight* in France, *we shall fight* on the seas and oceans . . . *we shall fight* on the beaches, *we shall fight* on the landing grounds, *we shall fight* in the fields and in the streets, *we shall fight* in the hills; we shall never surrender. . . .[33]

Martin Luther King, Jr., a leader of the US civil rights movement, was a superb orator, and frequently used repeating phrases to animate his audience. In a speech that he gave on the steps of the State Capitol in Montgomery, Alabama, on March 25, 1965 he declared in answer to his own question of long it will take for justice to prevail:

*How long? Not long,* because no lie can live forever. *How long? Not long,* because you shall reap what you sow. *How long? Not long.* . . .[34]

The repeated chanting of words and phrases in chorus is remarkably effective in political situations. Mexican farm workers in the 1960s continuously chanted "*Huelga, huelga, huelga,*" meaning "strike," as they marched, picketed, and demonstrated in their fight for higher wages and better living conditions.[35] They adopted the slogan "*Sí, se puede*" as their rallying cry, and its translation, "Yes, we can," was repeatedly chanted in chorus to spectacular effect during Barack Obama's 2008 presidential campaign. In a famous speech on January 8, 2008 in Nashua, New Hampshire, Obama declared:

*Yes, we can,* to justice and equality. *Yes, we can,* to opportunity and prosperity. *Yes, we can* heal this nation. *Yes, we can* repair this world. *Yes, we can.*[36]

**Speeches involving Repetition**

http://dianadeutsch.net/ch10ex05

In these speeches, which you can hear in the accompanying module, the repeating words and phrases are persuasive, not because repetition is necessary to drive home their meaning, but rather because the sound of a repeating phrase is mesmerizing, and serves to solidify and strengthen bonds that link people together.

Poetry makes extensive use of repetition—again, not to facilitate meaning, but rather to produce a mesmerizing effect. Take the end of Robert Frost's "Stopping by Woods on a Snowy Evening":

> The woods are lovely, dark, and deep,
> But I have promises to keep,
> And miles to go before I sleep,
> And miles to go before I sleep.[37]

These exceptions aside, repetition occurs far more commonly in music than in speech. When a mother sings a lullaby to her baby, the frequent repetition of tonal patterns and rhythms enables the baby to become familiar with the music, and this induces a sense of security. Consider children's songs, such as "Skip to My Lou," and "London Bridge Is Falling Down," which typically involve considerable repetition of the musical phrases (and also of the words). Hymns and other religious songs are typically composed of verses, with the same music set to different words in each verse. Chants, marches, and waltzes all require that the participants be familiar with the music—and music needs to be played repeatedly if participants are to sing, dance, march, stomp, or clap along with it together.

᷉

In addition to repetition, another effect is involved in the perceptual transformation from speech to song. Vowels that are produced by the human voice consist of harmonic series, and so give rise to perceived pitch.[38] The pitches of sung vowels are distinctly heard, and together they produce melodies. In contrast, the pitches of spoken vowels are much less salient, and they sound watered down. It appears that when we listen to the normal flow of speech, the neural circuitry that's responsible for pitch perception is inhibited to some extent, enabling us to focus our attention better on other characteristics of speech that are essential to conveying meaning; namely, consonants and vowels. We can assume, then, that exact repetition of spoken words causes this circuitry to become disinhibited, with the result that their pitches become more salient, and so more appropriate to song.

If a single vowel is presented repeatedly with very brief pauses between repetitions, after a short while it loses its vowel-like quality and instead sounds rather like a horn with a clear pitch. I came to realize the strength of this effect some years before I discovered the speech-to-song illusion. In the 1980s, I was working with my friend and collaborator Lee Ray on developing software to determine the average pitch of a vowel sound. I'd record a vowel into a computer, then Lee would chop off the initial and final consonants so that only the vowel remained. For example, I'd record the work "Mike," and Lee would chop off the "m" and the "k" sounds. Then the software would compute the average pitch of the vowel, and we would loop the vowel sound and listen to the looped version, with the result that we would hear the pitch distinctly. Then, as a rough check of the software, I would match the pitch of the looped vowel sound that I heard coming from the computer with a note I played on a synthesizer.

We kept repeating this sequence of events—record a vowel sound, loop it, decide what pitch we were hearing, check our decision by playing a note on a synthesizer – and we kept doing this for hours. Then suddenly I realized that working at the boundary between speech and song for such a long time had caused me to hear musical tones rather than speech sounds, and as a result I was having difficulty speaking! Alarmed, I turned to Lee and said—with difficulty—"I'm having trouble speaking—let's quit now," and Lee replied "I'm having trouble, too." So without another word, we packed up our equipment and left. And we never had a long session like that one again.

The speech-to-song illusion occurs in listeners regardless of musical training.[39] In addition to English, it occurs in phrases that are spoken in many languages, including German, Catalan, Portugese, French, Croatian, Hindi, Irish, Italian, and even the tone languages Mandarin and Thai. Utterances that are spoken in nontone languages that are more difficult to pronounce are associated with a stronger illusion, and the strength of the effect is reduced in tone languages.[40-42]

A group of us—Adam Tierney, Fred Dick, Marty Sereno, and I—carried out an experiment that supported the conjecture that pitch salience is involved in the speech-to-song illusion.[43] Through an extensive search of audiobooks, Adam uncovered further spoken phrases that came to be heard as sung following several repetitions. He also identified a second set of phrases that continued to be heard as spoken following repetition, and were matched with the first set on a number of characteristics. Subjects listened to both sets of phrases, with each phrase played repeatedly, while their brains were scanned using fMRI. We found that a network of brain regions in the temporal and parietal lobes that had previously been associated with pitch salience, perception of pitch patterns, and memory for pitch responded more strongly when the subjects were hearing a phrase as though sung rather than spoken. In contrast, no brain regions responded more strongly when the subjects were hearing a phrase as spoken rather than sung. So singing with words involves more brain circuitry than speaking in the absence of melody. On consideration, this is not surprising, as when we sing we need to process both the words and the music, rather than words alone.[44]

✑

Here's another line of reasoning: When songwriters set words to music, they combine the prosodic features of the lyrics with the melody, and create a rhythm in which strong syllables occur on strong musical beats. Now the linguistic rhythm as well as the pitch contour of the spoken phrase "sometimes behave so strangely" (as it occurs on my CD) are both very similar to those of a musical pattern, and this similarity could account in part for why the listeners' perceptions are transformed so smoothly from speech to song.

Memory factors are also involved here. If the sequence of pitches that forms a repeating phrase can be organized in terms of a familiar melody, then this familiarity can be used as an additional cue to treat the phrase as sung rather than spoken. The same is true of the temporal pattern formed by the repeating phrase. In the unlikely event that both the sequence of pitches and the temporal pattern coincide with those of familiar melodies, then together they provide even stronger cues that the phrase is being sung rather than spoken.

Here's my favorite explanation for why this particular spoken phrase "sometimes behave so strangely," as recorded on the CD, produces such a strong illusion. The basic pitch pattern forming this recorded phrase is very close to that of a phrase in the famous Westminster chimes. Furthermore, the rhythm of the phrase is essentially identical to that in the well-known Christmas song "Rudolph the Red-Nosed Reindeer." We must assume that there exists in the brain a database of well-remembered pitch patterns, and another database of well-remembered rhythms, and that we recognize songs by accessing these databases. Let's suppose, then, that the brain circuitry underlying memory for melodies recognizes the Westminster chimes, and that the brain circuitry underlying memory for rhythms recognizes "Rudolph the Red-Nosed Reindeer." These two matches, taken together and combined with the cues provided by repetition, lead to the conclusion that song rather than speech is being produced—so the brain mechanisms responsible for analyzing this pattern as song are invoked.

ʘ

We next continue to examine relationships between music and speech, including ways in which a person's language or dialect can influence his or her perception of music, and how musical experience can influence the perception of language. The strong relationships that have been revealed between these two forms of communication then lead us to enquire how present-day speech and music may have emerged in our evolutionary history.

# 11

## Speech and Music Intertwined

### CLUES TO THEIR ORIGINS

~⌐─────────────────────────────────────────────────

What we regard as the distinctive traits of song, are simply the traits of emotional speech
intensified and systematized . . . vocal music originally diverged from emotional speech in a
gradual, unobtrusive manner.

—HERBERT SPENCER, *On the Origin and Function of Music*[1]

THE SIXTEENTH-CENTURY MUSIC theorist and composer Vincenzo Galilei (the
father of Galileo) claimed that, for music to attain its full depth of expression,
it should have characteristics of speech—particularly the speech of someone experiencing a strong emotion. This argument strongly influenced the composer
Claudio Monteverdi, who incorporated phrases intermediate between speech and
song into his operas, such as *L'Orfeo*, in which emotions are expressed with extraordinary realism.

In the eighteenth century, the linguist Joshua Steele explored the relationship between music and speech in detail. In his treatise *An Essay Towards
Establishing the Melody and Measure of Speech to be Expressed and Perpetuated by
Peculiar Symbols*, [2] he proposed a modified form of musical notation to capture
the prosody of speech, and he observed that this scheme would provide a record
for future generations to determine the pitch, timing, and stress values in the
speech of his day. He acknowledged that differences exist between speech and
music—for example, he represented speech on an expanded musical staff that
was divided into steps smaller than a semitone, and he also pointed out that

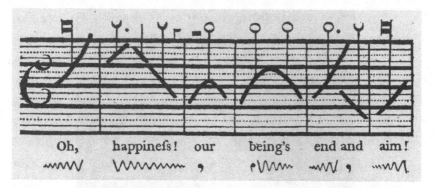

FIGURE 11.1. Transcription by Joshua Steele of a spoken phrase. From Steele (1779/1923, p. 13).

speech consists mainly of sliding tones rather than the discrete pitches that are typical of music.

To document the pitches that he heard from a speaking voice, Steele took a bass viol, and marked pitch values in quarter-tone steps on paper attached to the side of the board. He then bowed across a string, which he fingered in accordance with the pitches produced by his own speaking voice, and he designated the pitches that he heard on this grid. Next, he transferred this information onto a modified musical staff, using curved lines to indicate pitches and their trajectories. Figure 11.1 gives an example of Steele's transcription of a spoken phrase in his quasi-musical notation.

In the nineteenth century, Charles Darwin (Figure 11.2) moved beyond acknowledging similarities between music and speech to hypothesize that the first protolanguage was musical, and that it was used primarily for courtship. In his seminal book, *The Descent of Man, and Selection in Relation to Sex*[3] he wrote:

> It appears probable that the progenitors of man, either the males or females of both sexes, before acquiring the power of expressing their mutual love in articulate language, endeavored to charm each other with musical notes and rhythm. (p. 639.)

In support of his argument, Darwin made comparison with birdsong, which is most often produced by males during the breeding season. As he wrote:

> The true song . . . of most birds and various strange cries are chiefly uttered during the breeding-season, and serve as a charm, or merely as a call-note, to the other sex. (p. 417.)

FIGURE 11.2.  Charles Darwin.

And later:

> Unless the females were able to appreciate such sounds and were excited and
> charmed by them, the persevering efforts of males, and the complex struc-
> tures often possessed by them alone, would be useless; and this is impossible
> to believe. (pp. 635–36.)

Darwin then advanced a similar argument concerning human music. He noted
that music occurs in all cultures, develops spontaneously in children, and gives
rise to strong emotions. As he wrote:

> All these facts with respect to music and impassioned speech become intelli-
> gible to a certain extent, if we may assume that musical tones and rhythm were
> used by our half-human ancestors, during the season of courtship. (p. 638)

Following the publication of *The Descent of Man*, Herbert Spencer argued in a
postscript to his essay *On the Origin and Function of Music* that in proposing that

music first arose in the context of courtship, Darwin had interpreted its origins too narrowly. Spencer maintained that music instead developed from vocalizations that were produced in a wide range of emotional states, such as joy, triumph, melancholy, grief, and rage, in addition to sexual excitement.

Similarly to Darwin, Spencer bolstered his case by considering birdsong, which, he argued, occurs not only in courtship but also in other situations. In addition, he cited examples of human primitive music that occur in venues that are unrelated to courtship, such as rituals, work songs, laments, and choral singing. But he agreed with Darwin's claim that both speech and music have their roots in a common protolanguage.

Early in the twentieth century, the Danish linguist Otto Jespersen, in his book *Language: Its Nature, Development, and Origin*, also proposed that song and speech emerged from a primitive form of communication that had elements of both. In this context, he proposed that tone language preceded contemporary nontone languages:

> Men sang out their feelings long before they were able to speak their thoughts . . . When we say that speech originated in song, what we mean is merely that our comparatively monotonous spoken language and our highly developed vocal music are differentiations of primitive utterances, which had more of them of the latter than of the former.[4] (p. 436.)

Jespersen contended that spoken language initially consisted of long, inseparable utterances, and that these were divided into progressively smaller units with the passage of time, leading to the words we use today. Recently, Alison Wray proposed in detail that protolanguage was holistic, and consisted of lengthy utterances that had specific functional meanings, but lacked internal structure.[5] As she pointed out, we still use such utterances in certain situations. For example, *haudjudu* means "I politely acknowledge the event of our initial meeting," and *stikaroun* means "kindly remain in the vicinity." On the other hand, if only lengthy utterances with specific meanings like these were available to us, then the attempt to convey detailed information would impose an unacceptable load on memory. In contrast, contemporary language, which is built from a lexicon of small words together with rules by which they can be combined, enables us to convey complex information much more easily.

Both speech and music have syntax, in which a relatively small set of basic elements are sampled and organized in accordance with certain rules so as to create hierarchical structures. In linguistic syntax, words combine to form phrases, and these join to form longer phrases, which in turn combine to form sentences. In music, notes combine to form motifs, which in turn are linked to form phrases, which themselves join to form longer phrases; and so on. A hierarchical form of organization is important to both speech and music, as it allows a small number of elements to be combined in a potentially infinite number of ways, so enabling us to express creative and original thoughts.

There are critical differences between linguistic and musical syntax, however. Language is based on the conventional pairing of sound with meaning. For example, hearing the word *cat* invokes the concept of a cat, since we've learned from experience to link the sound with the concept. So we must have stored in our brains a lexicon of words together with the concepts they represent, and we must also have stored a set of rules for combining words together, so as to infer relationships between the concepts.

Words represent concepts that are grouped into categories termed *parts of speech*—for example, roughly speaking, nouns stand for things, and verbs stand for actions—so the rules for linking words together must be tailored to the verbal categories involved. There's no counterpart in music to relationships between sounds and concepts, or between sounds and parts of speech. In addition, statements in language can have truth values—for example, if I say, "There is an elephant walking through the room," this statement is untrue (I could be lying or hallucinating). Yet there is nothing corresponding to truth values in music. Similarly, there is no musical counterpart to negation—as if, for example, I say, "Joe has not arrived."

Similarly, musical syntax involves basic properties that have no counterpart in language. As shown in the Appendix, Western tonal music has twelve possible "pitch classes"—the notes [C, C♯, D, D♯, E, F, F♯, G, G♯, A, A♯, B], which repeat at octave intervals. Subsets of these notes called *scales* exist at a higher level—for example, the C major scale is composed of the notes [C, D, E, F, G, A, B]. At an even higher level there are subsets of notes taken from a scale—for example, the triad beginning on the note C, taken from the C major scale, consists of the notes [C, E, G]. Phrases in much Western tonal music consist of notes that are arranged as hierarchies that correspond to these different levels.

◌▵

The aspect of present-day English speech that's most closely related to music is *prosody*, a term used to describe the sound patterns of speech; that is, variations in pitch, tempo, timing patterns, loudness, and sound quality—these emphasize the structure of phrases, and also serve to convey emotional tone.[6] Like syntactic structures, prosodic structures are hierarchical—syllables combine to form feet, and these combine to form words, which themselves combine to form phrases.

Prosody helps us to decode and so understand the flow of speech. For example, in southern British English, questions usually end with an upward pitch movement, while statements end with a movement in the downward direction. Boundaries between phrases are generally signaled by pauses, and the endings of phrases tend to be characterized by words of longer duration, which are often spoken at lower pitches.

Where whole phrases are considered, prosody can contribute to meaning. Important words in a sentence tend to be stressed, and the overall contour of a sentence can vary depending upon whether it's a statement, a question, or an exclamation. When I was an undergraduate at Oxford, I attended a series of lectures by the great philosopher J. L. Austin. In one of these lectures, he showed how emphasizing a word in a sentence can alter its entire meaning. As an illustration, he pointed out that the meaning of the sentence "There is a booming bittern at the bottom of the garden"[7] varies depending on which word is stressed:

> There is a BOOMING bittern at the bottom of the garden
> There is a booming BITTERN at the bottom of the garden
> There is a booming bittern at the BOTTOM of the garden
> There is a booming bittern at the bottom of the GARDEN

Not only does the pitch contour of a sentence provide information about its intended meaning, but it also varies depending on the language or dialect that's being spoken. As mentioned earlier, in southern British English a sentence ending on a higher pitch is generally a question, whereas one ending on a lower pitch generally signals a statement. However, this isn't true of all English dialects. Young women in southern California, where I teach, often speak to me sounding rather like this: "Professor Deutsch—my name is Melissa Jones? I'm writing a paper on music perception?" But these sentences are intended as statements—not as questions. Scholars have dubbed this form of speech *Uptalk* (sometimes referred to as "Valley Girl Speak"). When I demonstrated this to my students at UCSD, they found my imitation of Uptalk hilarious. But when I delivered a statement ending with a downward sweep, appropriate to much speech in England, they became distinctly uneasy—perhaps this form of intonation appeared patronizing to them.

Patterns of intonation vary widely within England, depending on where the speaker grew up, and on his or her social class. The playwright George Bernard Shaw was intrigued by intonation patterns. In Shaw's time, people needed to speak in an upper class (or at least a middle class) accent if they were to be hired for certain jobs. In Shaw's play *Pygmalion*, the phonetician Henry Higgins speaks disparagingly about a flower girl:

> You see this creature with her kerbstone English: the English that will keep her in the gutter to the end of her days. Well, sir, in three months I could pass that girl off as a duchess at an ambassador's garden party.[8]

Meter—the alternation of strong and weak beats that are equally spaced in time—occurs in music, and to a lesser extent in speech. In both cases, meter can be described in terms of a hierarchy of levels, such that the stronger the beats, the closer they are to the top of the hierarchy. As shown in the Appendix, a metrical hierarchy is often represented as a grid containing horizontal rows of symbols, and the more prominent the events (syllables in speech; notes in music), the higher up they are represented.

While metrical structure is generally very important in music, and is also important in poetry, it exists rather loosely in conversational speech. This is analogous to the difference in pitch structure between speech and music. For the most part, pitches that are performed in music are constrained by the patterns of notes on the printed score; however, pitch patterns in speech are less constrained by the spoken words (at least in nontone languages such as English). Moreover, patterns of timing in music can be very strict, yet they are more relaxed in conversational speech.

Prosody is also important in communicating emotion. In a meta-analysis of studies on vocalization patterns in speech and music, the psychologists Patrick Juslin and Petri Laukka found that these two forms of expression showed a remarkable similarity with respect to profiles conveying happiness, sadness, tenderness, and fear.[9] For example, in both cases happiness tends to be signaled by fast tempi, medium-to-high amplitudes, and high, rising pitches; sadness by slow tempi, low amplitudes, and low, falling pitches. So, following Herbert Spencer's proposal, these authors argued that music imitates the vocal characteristics of speakers who are possessed of certain emotions. Taking this further, we can surmise that the neural circuits involved in emotional response to music may have evolved in the context of a protolanguage that had elements of both music and speech.

ᎧᏋ

Many studies have documented a close relationship between speech and music. By the time babies are born, they are already familiar with the melodies of their mother's speech. Sound recordings taken from inside the womb at the onset of labor have shown that the mother's speech sounds can be loudly heard.[10] However, the sounds are filtered through the body tissues, so that the high frequencies—important for identifying the meanings of words—have been muffled, while the musical (or prosodic) characteristics of speech – pitch contours, variations in loudness, rhythmic patterns, and tempo – are well preserved. And during the last part of pregnancy, babies in the womb respond to their mothers' speech by decreasing their heart rates,[11] indicating that even before babies are born, their mother's speech is important to them.

Newborn infants prefer to listen to recordings of the voices of their mothers, rather than those of other women.[12] They also prefer to listen to their mothers' low-pass filtered voices over other women's voices.[13] And very young babies prefer to listen to recordings of speech in their native language rather than in a language that's foreign to them.[14] So infants prefer most to listen to their mothers' speech, next to a different woman speaking the same language, and least of all to speech in another language. The preference of infants for the music of their mothers' speech must contribute strongly to the formation of bonds between mothers and their babies—bonds that are important to the infant's survival.

The familiarity of newborns with the pitch contours of their mothers' speech is also apparent in their vocalizations. The cries of newborn babies tend first to rise and then to fall in pitch. When the cries of neonates who had been born into families who spoke either French or German were recorded, the wails of the French babies consisted mostly of the rising portion, while the descending portion was more prominent in the cries of the German babies. Now, rising pitches occur more commonly in French speech, whereas in German speech falling pitches are more common. So it appears that the wails of newborn babies possess some of the musical qualities of the speech to which they had been exposed before birth.[15]

The melody of speech is important for communication between mothers and their infants in another way also. The sounds produced by parents speaking to their babies are strikingly different from those of normal adult speech. They involve shorter phrases, higher pitches, larger pitch ranges, slower tempi, longer pauses, and abundant repetition. These exaggerated speech patterns, termed *motherese*, occur in many languages around the world. Mothers use falling pitch contours when calming a distressed baby, and rising contours to engage the baby's attention. They use steeply rising and then falling pitch contours to express approval, as in "Go-o-o-d girl!" They express disapproval less frequently, but when they do so, the mothers use low, staccato voices, as when saying "Don't do that!" The babies respond appropriately, even though they don't understand the meanings of the words they hear—they smile when they hear "approvals," and become subdued or cry when they hear "prohibitions."

In a clever study by Ann Fernald at Stanford University, when five-month-old infants from English-speaking families listened to "approval" and "prohibition" phrases that were spoken in nonsense English, German, Italian, and English motherese, they responded with the appropriate emotion to utterances in all these languages, even though the speech was nonsensical to them. So even though babies at this age can't comprehend the meanings of words, they can still understand the intention conveyed by the melody of speech.[16]

I have two children, and when they were babies I would instinctively talk to them in motherese. But later, when I read these descriptions of motherese, I wondered whether, in general, babies react to this form of speech as powerfully as was described. So one day when I was shopping in a supermarket and saw a mother with a baby in a stroller, I approached the baby and said, in my best motherese voice that glided steeply upward and then downward, "Hel - lo - o!" To my surprise I was rewarded with a huge smile, and the baby continued to look at me intently for a long while, her eyes following me as I walked away! Fortunately the mother didn't object to my intrusion.

The slow, exaggerated patterns of motherese also help infants to take their first steps in learning to talk. The slow tempi, pauses between words and phrases, highlighting of important words, and exaggerations of pitch patterns make it easier for them to interpret the syntactic structure of their native language. In one study, babies around seven months of age were presented with nonsense sentences that were spoken either in normal adult speech or in motherese. In both cases, the statistical structure of the speech provided the only cue to word boundaries. Even though the speech was nonsensical, the babies were able to detect words that were embedded in the motherese sentences, though they couldn't do this for the sentences in normal speech.[17]

Although people have an inborn tendency to detect and respond to the melody of speech, this ability can be enhanced by music lessons. In one experiment, eight-year-old children who hadn't received musical training were divided into two groups. Those in the first group were given music lessons for six months, while those in the second group were given painting lessons. Before and after this six-month period, the children were presented with recorded sentences, in some of which the last word had been raised slightly in pitch so as to sound somewhat incongruent with the rest of the sentence. The children were asked to identify the altered sentences. Although the performance of the two groups didn't differ initially, at the end of the six months, the children who'd taken the music lessons surpassed the others on this test. These findings indicate that musical training can help children grasp the prosody, and so the intended meaning, of spoken utterances.[18]

The fact that prosody is involved in conveying emotion leads to the surmise that recognition of emotions in speech would also be enhanced by music lessons. In a study by William Thompson and his colleagues at the University of Toronto,[19] six-year-olds were given keyboard lessons for a year, and they were then asked to detect emotions that were expressed in spoken sentences. In identifying whether the sentences expressed fear or anger, these children performed better than did other children who'd received no keyboard lessons during this period, and this was true even for sentences that were spoken in an unfamiliar language.

Studies on adults have led to similar conclusions. In one study, French musicians and non-musicians were played sentences that were spoken in Portugese. In some of the sentences, the final words were incongruous in pitch, and the musicians outperformed the non-musicians in detecting the incongruities.[20] In another study, Portuguese speakers listened to short sentences that expressed six different emotions—fear, anger, disgust, sadness, happiness, and surprise. Listeners with musical training outperformed the untrained listeners in recognizing all these emotions.[21] In a related study, congenitally amusic individuals were found to be significantly worse than matched controls in their ability to understand the emotions conveyed by prosodic cues.[22]

Neurophysiological studies have pointed to an advantage for musically trained listeners in acquiring lexical tones. The neuroscientists Patrick Wong and Nina Kraus at Northwestern University, together with their colleagues, studied the auditory brainstem responses of English speakers to Mandarin speech sounds. The responses of the musically trained speakers were much stronger—and the earlier in life they had begun training, and the longer the training had continued, the stronger were the brainstem responses.[23]

The reasons for relationships between childhood musical training and adult speech perception are still being debated, however. Children who take music lessons are likely to have a higher musical aptitude than others, and this could in turn be associated with enhanced speech processing. We therefore need to consider carefully the role of innate factors in evaluating the relationship between musical training and speech perception.[24]

ᑲ

Not only does music influence the way we perceive speech, but the speech we hear also influences our perception of music. The tritone paradox—a musical illusion that we explored in Chapter 5—provides a striking example. This illusion is produced by tones that are generated by computer so that their note names (C, C♯, D, and so on) are clearly defined, but the octave in which they are placed is ambiguous. Two such tones that are related by a half-octave (termed a tritone) are played in succession, and the listener decides whether the pattern ascends or descends in pitch. As described in Chapter 5, such judgments vary depending on the language or dialect to which the listener has been exposed, particularly in childhood. For example, I found that listeners who had been raised in California and those who had been raised in the south of England tended to hear this illusion in roughly opposite ways—in general when the Californians tended to hear a tritone pattern as ascending the southern English group tended to hear the same pattern as descending; and vice versa. In another study, my colleagues and I found that listeners who had been born in Vietnam heard the tritone paradox quite differently from native speakers of English who had been born and grown up in California, and whose parents had also been born and grown up in California. Perception of the tritone paradox also correlates with the pitch range of the listener's speaking voice, and this in turn varies depending on his or her language or dialect. So this illusion reflects a direct connection between how a person speaks and the way he or she hears this musical pattern.

A further influence of language on music perception concerns absolute pitch—the ability to name the pitch of a note (C, C♯, D, etc.) in the absence of a reference note. As described in Chapter 6, my colleagues and I found that, among students in several music conservatories, the prevalence of absolute pitch was far higher among speakers of a tone language such as Mandarin, Cantonese, and Vietnamese than among speakers of a nontone language such as English. In another study, we found that speakers of tone languages were remarkably consistent in the pitches with which they pronounced lists of words in their native language—this indicated that they had established fine-grained pitch templates for the perception and production of words in their language. Tone language speakers process musical pitches better than do nontone language speakers in other ways also—they imitate the pitches of tones and detect pitch changes in tones more accurately, and are better able to discriminate musical intervals, compared with speakers of nontone language such as English. This tone language advantage even occurs in young children: Mandarin-speaking children of ages 3–5 perform better than English-speaking children of the same ages in discriminating pitch contours.[25]

The advantage to pitch processing of speaking a tone language has also been shown in electrophysiological studies. In one study, researchers examined the auditory brainstem response to Mandarin words, and they found that Mandarin speakers exhibited stronger pitch representation than did English speakers.[26] In another study, Cantonese-speaking non-musicians and English-speaking musicians outperformed English-speaking non-musicians, both on various auditory tasks and also on general cognitive tasks involving short term memory and general intelligence.[27] These findings point to an influence of tone language as well as musical training on auditory function and on general cognitive ability.

The language we speak also affects the way we process timing in music. In Finnish, the meaning of a word can differ depending on the duration of a syllable in the word. For example, the Finnish word *sika* means "pig," whereas *siika* means "whitefish." Accordingly, Finnish speakers surpass speakers of French in discriminating between musical sounds that differ only in duration.[28] Considering the relationship between speech and the timing characteristics of music from a broader perspective, musicians have surmised for centuries that instrumental music might reflect the prosody of the composer's native language. Aniruddh Patel and Joseph Daniele examined this hypothesis by considering the contrast between durations of successive elements of speech in different languages.[29] They compared British English, where this contrast tends to be large, with standard French, where it tends to be smaller. Using Barlow and Morgenstern's *Dictionary of Musical Themes*, the authors examined themes from classical instrumental music produced by English and French composers who were active in the late nineteenth and early twentieth centuries. They found that the amount of contrast between the durations of successive tones in the music of these two countries differed in the same direction as the durational contrast in the spoken utterances. So this study indicates that when composers produce music, they unconsciously draw on the timing patterns of the speech they have been hearing since childhood.[30]

In a related study, John Iversen and colleagues presented Japanese and English listeners with patterns of tones that consisted of repeating sequences either of the "long-short" type, or of the "short-long" type. In accordance with the speech patterns of their language, the Japanese listeners preferred the "long-short" sequences, while the English listeners preferred those that were "short-long."[31]

◌

While prosody is of particular importance in nontone languages such as English, pitch assumes even more importance in tone languages—that is, languages that use pitch to convey the meanings of individual words. Tone languages are spoken by well over a billion people. In languages such as Mandarin, Cantonese, Vietnamese, and Thai, a word takes on an entirely different meaning depending on the pitch (or pitches) in which it is spoken. In English we can say a word such as *dog* in many different pitches, without changing its essential meaning. But in tone languages, the meaning of a word differs depending on its pitch and pitch contour.

If you try to speak in a tone language without paying attention to the pitches of the individual words and syllables, you risk being seriously misunderstood. Yingxi

Shen, then a graduate student who was visiting me from Guangzhou (Canton) in China, gave me the following example from Mandarin. Suppose you intend to say: "I want to ask you." This is *"wo xiang wen ni."*[32] If you mispronounce *wen* so that it's spoken in the wrong tone, you are saying "I want to kiss you"—which could be embarrassing.

The English missionary John F. Carrington described another embarrassing source of confusion. The drum language Kele (a language in Zaïre) has two tones—high and low—and you need to be careful which tone you use for each syllable. For example, altering the tones in *"alambaka boili"* can transform the meaning of this phrase from "he watched the riverbank" into "he boiled his mother-in-law"!

To give you an idea of the importance of the pitches of words in a tone language, here's an informal experiment that Yingxi and I carried out when she was visiting my lab at UCSD. First, Yingxi recorded some Cantonese words into a computer, and this enabled us to analyze their pitches so as to obtain an informal estimate of the range of her Cantonese-speaking voice. Yingxi then chose the word *faan* for our experiment, since it has three flat tones—meaning "double" in the high flat tone, "to spread" in the mid flat tone, and "prisoner" in the low flat tone. Then, using a computer, I transposed the recording of *faan* into nine different pitches (while keeping the timing of the word constant), and so produced words that varied in semitone steps within the range of Yingxi's speaking voice. In this way, we ended up with a bank of words ranging from A3 (three semitones below middle C) as the highest pitch, down to C♯3 (almost an octave below middle C) as the lowest. In our experiment, I first played *faan* in one of these pitches, and Yingxi gave me the meaning of the word she heard. Then I played *faan* in another of these pitches, and she gave me the meaning of that word; and so on for all the nine pitches we used. Here's what happened when we began with the highest note (A3) and moved down in semitone steps to the lowest note (C♯3).

| Pitch of *faan* | Yingxi's response |
| --- | --- |
| A3 | "It's saying 'double'" |
| G♯3 | "'Double' again" |
| G3 | "Now it's saying 'to spread'" |
| F♯3 | "'To spread' again" |
| F3 | "'To spread' again" |
| E3 | "'To spread'" |
| D♯3 | "Now it's saying 'prisoner'" |
| D3 | "'Prisoner' again" |
| C♯3 | "'Prisoner' again" |

The idea that the pitch of a word can be important to its meaning often appears strange to speakers of English, but it's natural for tone language speakers. Not all monosyllabic words in Mandarin, Cantonese, or Vietnamese have meanings in all the tones of the language; a word can have several different meanings when it's spoken in certain tones, and yet have no meaning in other tones. This is analogous to the way we use vowels in English. Take the word B*T. This has many different meanings depending on the vowel it contains. "BAT," "BATE," "BEAT," "BIT," "BITE," "BUT," "BOUGHT," "BOAT," "BET," and "BOOT" all have meanings. But then take the word P*G for comparison. "PEG," "PIG," and "PUG" have meanings, but "PAG," "PAIG," "PEEG," "PIGE," "POG," "POAG," and "POOG" are nonsensical.

Things get even get stranger when we consider that a language such as Mandarin has a large number of homophones (words that have more than one meaning), including words that are spoken in the same tone. Yuen Ren Chao, who was a distinguished linguist, poet, and composer, provided an amazing example. He wrote a story in ancient Mandarin in which the word *shi* is used about a hundred times, invoking the different tones and making use of its many homophones.

The story is titled "The Lion-Eating Poet in the Stone Den" and although it's rather peculiar, it does make sense. Here it is in Pinyin (the standard phonetic system for transcribing Mandarin pronounciations of Chinese characters using the Latin alphabet):

Shī Shì shí shī shǐ »
Shíshì shīshì Shī Shì, shì shī, shì shí shí shī.
Shì shíshí shì shì shì shī.
Shí shí, shì shí shī shì shì.
Shì shí, shì Shī Shì shì shì.
Shì shì shì shí shī, shì shǐ shì, shǐ shì shí shī shìshì.
Shì shí shì shí shī shī, shì shíshì.
Shíshì shì, Shì shǐ shì shì shíshì.[33]

And here's the translation:

Lion-Eating Poet in the Stone Den
In a stone den was a poet with the family name Shi, who was a lion addict,
    and had resolved to eat ten lions.
He often went to the market to look for lions.

At ten o'clock, ten lions had just arrived at the market.

At that time, Shi had just arrived at the market.

He saw those ten lions, and using his trusty arrows, caused the ten lions
  to die.

He brought the corpses of the ten lions to the stone den.

The stone den was damp. He asked his servants to wipe it.

After the stone den was wiped, he tried to eat those ten lions.

When he ate, he realized that these ten lions were in fact ten stone lion
  corpses.

Try to explain this matter.

Due to their large number of homophones, many monosyllabic words in Mandarin and Cantonese are ambiguous when spoken in isolation, and take on meanings only in conjunction with other words. In consequence, the meaning of a word frequently needs to be disambiguated by adding context. English also has homophones, and you can play a similar—though less elaborate—trick with some of these. Here's a sentence that's composed of a single word that's repeated seven times—it has a meaning, though it may take you a few minutes to figure it out:

*Police police police police police police police.*

To make sense of this sentence, consider that "police police" could be a name for people who supervise the police. Then the "police police police" supervise the "police police." The full sentence can then be worked out from this.[34]

The connection between tone language and music is so strong that a naturally spoken phrase in a tone language can be conveyed on a musical instrument. The ehru—an ancient two-stringed bowed musical instrument that sounds somewhat like a cello—works well for this. Sometimes music hall artists in China will play a phrase or two on the ehru, and the delighted audience understands the intended verbal meaning. "*Nin hao*," meaning "hello" in Mandarin, works very well. I asked my friend and colleague, Professor Xiaonuo Li, who heads the research branch of the Shanghai Conservatory of Music, if she would ask a performer on the ehru to illustrate this effect. She arranged for a student, Ziwen Chen, to play the phrases "Hello," "I can speak," "I love you," and even "Nice weather! It is sunny and blue" as it is spoken in Mandarin. On listening to these musical phrases, Mandarin speakers understood their verbal meanings. More rarely, musical performers sometimes use similar tricks when they imitate intonation patterns in a nontone language. For example Jimi Hendrix was famous for saying "thank you" to his audiences by playing two glissandi on his guitar.[35]

℘

Earlier we explored the issue of whether speech and song should be considered the function of separate modules, or whether they both reflect a system that should be regarded as a monolithic whole. It's clear that neither of these extreme views is correct. Speech and music both involve the vocal/auditory channel of our nervous system. They also involve the processing of pitch structure, timing, tempo, metrical structure, sound quality, and loudness, though in different ways. In addition, they are also both hierarchically structured, and both consist of a finite number of discrete elements (words or tones) that can be combined in an infinite number of ways so as to create larger structures in the form of phrases. And they both involve general systems, such as attention, memory, emotion, and motor control.

Yet speech and music differ in important respects. Speech conveys propositional meaning—it informs the listener about the state of the world. This is achieved using a lexicon of words together with the concepts they represent. Words map onto objects and events—for example, the word *dog* represents a dog, and the word *run* designates the event of running. A set of rules (incorporated as syntax) combines these words so as to enable more complex ideas to be expressed. In music, tones are also combined in accordance with a set of rules; however, these rules are entirely different from those that are used in language (see Appendix). Also music heavily involves pitch structures, rhythm, and meter—in contrast, with some exceptions, speech doesn't have the regular timing structure that's characteristic of music. And importantly, repetition is a vital element of music, whereas with some exceptions, such as poetry and oratory, it is less vital in speech.

Speech and music also differ substantially in their functions. While speech serves primarily to inform the listener about the world, music is employed in many ways to modulate feelings and emotions. Mothers sing to their babies to soothe them and elevate their mood. Songs and instrumental music serve to promote cohesiveness among members of a group. Music occurs in religious ceremonies, rituals, and political rallies; it instills courage in soldiers during battle; and its rhythmic structure encourages people to exert themselves in manual labor. Love songs are used in courtship, as is evident from the content of many popular songs.

Finally, we revisit the question of how speech and music evolved. Given their similarities, we can discount the notion that they evolved separately and in parallel. As argued by Tecumseh Fitch, both speech and music may have had their origins in a vocal generative system, often called *musical protolanguage*, in which a small set of sound elements (phonemes) which are themselves meaningless are combined into larger structures. Such a system would be able to generate a huge repertoire of sound combinations. Larger units can then be formed from these basic units, and these can in turn form even larger units, and so on. Such a generative process would be hierarchical, with syllables, phonological words (or notes), and phonological phrases being built from the basic units. Today, such a hierarchical system is used in speech, but meaningfulness (semantics) is also very important here. In comparison, music is the function of a generative system that can produce an infinite set of meaningless structures, by hierarchically combining the basic elements (mostly notes) into increasingly larger musical phrases. So we can surmise that the generative aspect of phonology might have arisen before meaningfulness was involved in speech.

Darwin pondered deeply over the question of how our sophisticated communication system would have evolved from music-like sounds, and was unable to come up with an answer that he considered truly satisfactory. Nevertheless he wrote:

> It is not the mere articulation which is our distinguishing character, for parrots and other birds possess this power. Nor is it the mere capacity of connecting definite sounds with definite ideas; for it is certain that some parrots, which have been taught to speak, connect unerringly words with things, and persons with events. The lower animals differ from man solely in his almost infinitely larger power of associating together the most diversified sounds and ideas; and this obviously depends on the high development of his mental powers. (pp. 107–8.) [36]

Twentieth-century writers have also argued for a musical protolanguage view of the evolution of speech and music. These include Steven Mithen, who in his book *The Singing Neanderthals* described musical protolanguage with an acronym ("Hmmmm": indicating "holistic," "manipulative," "multi-modal," and "musical").[37] Steven Brown, another proponent of this view, coined the term *musilanguage model* to describe it.[38] He suggested (as had Jespersen) that early man used tone language, pointing out that this is true of the majority of the world's languages today. This suggestion appears very plausible, as pitch is particularly useful for the transmission of information—and it's noteworthy that even now, tone language is spoken by a substantial proportion of the world's population.

There are still many unanswered questions concerning the relationship between speech and music, particularly considering their molecular and genetic bases, and many conflicting views that need to be reconciled. However, given the research findings so far obtained, we should expect to see considerable progress in understanding this relationship in the years ahead.

# Conclusion

IN THIS BOOK, we have been examining a number of characteristics of sound perception and memory, and we have focused particularly on illusions. In the past, illusions of sound perception have tended to be dismissed as entertaining anomalies; yet as explored here, they reveal important characteristics of the hearing mechanism that might otherwise have gone unrecognized.

The illusions explored here show that there can be remarkable differences between listeners in how they perceive music. Some of these discrepancies are related to variations in brain organization, as reflected in differences between right-handers and left-handers in how the illusions are perceived. In the case of another illusion—the tritone paradox—perceptual variations are instead related to the listeners' languages or dialects. Taken together, these illusions show that both innate brain organization and environmental factors exert strong influences on what we hear.

Another way in which the illusions explored here shed light on the hearing system involves illusory conjunctions. When we hear a musical tone, we attribute a pitch, a loudness, a timbre, and we hear the tone as arising from a particular location in space. So we perceive each tone as a bundle of attribute values. It's generally assumed that this bundle reflects the characteristics and location of the emitted sound. However, in the octave and scale illusions it is shown that when several sequences of tones arise simultaneously from different regions of space, the bundles of attribute values can fragment and recombine incorrectly.

The phantom words illusion is also the product of illusory conjunctions. Here, components of one syllable that is emanating from one spatial location are perceptually attached to components of a different syllable that is emanating from another location. As a result, we obtain the illusion of an entirely different syllable that appears to be emanating from either one location or the other. Musical hallucinations also involve illusory conjunctions. Here, a segment of music may be "heard" correctly, with the exception of a particular attribute of the sound. For example, the segment may be "heard" as played in a different pitch range, tempo, loudness, or timbre.

Illusory conjunctions have important implications for our understanding of the overall organization of the auditory system. They lead to the conclusion that, in the normal way, the process of analyzing sound includes the operation of specialized modules that are each configured to analyze a particular attribute, and that we combine the outputs of these modules so as to obtain an integrated percept. This process usually leads us to perceive sounds correctly, but it breaks down under certain circumstances. The question of how we generally combine attribute values correctly provides an important challenge for future researchers.

Another thread that runs through the book involves the relationship between speech and music—an issue that has been debated for centuries. Recently, some researchers have argued that these two forms of communication involve different physical properties of sound, that they are distinct in their functions, and have different neuroanatomical substrates. Other researchers have argued, in contrast, that both forms of communication are subserved by a general system that is largely distributed across the brain. They contend that music and speech interact within this system, and that this is why the speech we hear is influenced by its musical qualities, and that the way we hear music is influenced by our language or dialect.

The speech-to-song illusion has caused a rethinking of both these viewpoints. Here a spoken phrase is made to be heard as sung, without altering its physical properties in any way, but simply by repeating it over and over again. This illusion has led me to conclude that processing speech and music involves both specialized modules—such as those subserving melody or rhythm in music, and semantics or syntax in speech—and also general factors. When a sound pattern is presented, a central executive in the brain determines, based on a number of cues—including repetition, context, and memory, together with the physical characteristics of the sounds themselves – whether it should be analyzed as speech or as music, and it transmits the information to be analyzed to the appropriate modules. From another perspective, earworms and musical hallucinations highlight the vital role played by the inner workings of our musical minds in determining what we hear.

The influence of unconscious inference (or top-down processing) on our perception of sound is also considered in detail in the book. As shown by the illusions that are explored, it plays a very important role in how we hear speech and music. For example, we can apprehend the "mysterious melody" described in the Introduction, but only when we already know what to listen for. As another example, the scale illusion and its variants are based in part on our knowledge and expectations concerning sound sequences. When we listen to the scale illusion, our brains reject the correct but improbable interpretation that two sources are producing sequences of tones that leap around in pitch. Instead, they create the illusion of two smooth melodies, each in a different pitch range. Furthermore, based on our knowledge and experience of sound sequences, most people hear the higher melody as coming from one source, and the lower melody as coming from a different source. The tritone paradox also provides an example of top-down processing, since we hear these ambiguous tritones as ascending or descending in pitch depending on our language or dialect, and so on our knowledge and expectations about sounds. The phantom words illusion displays a strong influence of unconscious inference on the perception of speech. It shows that in listening to ambiguous words and phrases, we draw on our knowledge and experience of speech and language, and the words we hear are also influenced by our state of mind, our memories, and our expectations.

In addition, extremes of musical ability are addressed. These are shown to be the product of the type of environment to which the listener has been exposed, as well as innate factors. As an example of environmental influence, the prevalence of absolute pitch is much higher among speakers of the tone language Mandarin than among speakers of English. At the other extreme, a small proportion of the population are tone deaf, and here there is a genetic component.

We have been exploring illusions, perceptual conundrums, and huge individual differences in the way people hear even simple musical patterns. These extraordinary phenomena lead us to conclude that the hearing mechanism did not arise as a coherent, integrated whole, but rather developed as a set of different, though interconnected, mechanisms, each with its own form of neural computation. Much remains to be learned, and I anticipate that the next couple of decades will take us much further in understanding the mechanisms that underlie the behavior of this remarkable, and ultimately highly successful, system.

*Appendix*

SEQUENTIAL SOUND PATTERNS IN MUSIC AND SPEECH

Sequential sound patterns in music and speech are structured in ways that are similar in some respects but different in others. Here we describe some ways in which we perceive tones and the relationships between them in Western tonal music, and we briefly compare these with how we represent patterns of words in language.

Musical instruments and the human voice produce complex tones; these consist of a combination of many pure tones (sine waves), termed *partials*. The tones produced by many instruments and the human voice are *harmonic complex tones*. The frequencies of their partials are whole-number multiples of the lowest frequency, termed the *fundamental frequency*. For example, if the lowest frequency is 100 Hz, the harmonic complex tone might also include tones at frequencies 200 Hz, 300 Hz, 400 Hz, 500 Hz, 600 Hz, and so on.

When a harmonic complex tone is sounded, the pitch that is heard corresponds to that of the fundamental frequency. Remarkably, this often holds even when the fundamental frequency is missing. For example, a tone that consists of harmonics at 200 Hz, 300 Hz, 400 Hz, 500 Hz, and 600 Hz is perceived as having a pitch that corresponds to 100 Hz. This has important practical consequences. For example, in the transmission of sound over landline telephones, all the frequencies below 300 Hz are absent. Yet the pitches of vowels produced by human voices, particularly those of men, generally fall below 300 Hz, so one might expect that most of these would not come through the phone lines. Yet because a sufficient number of harmonics above 300 Hz are present in the vowels, we infer the fundamental frequencies from these, and so hear the pitches of the transmitted vowels without difficulty.[1]

A characteristic of music in all known cultures is termed *octave equivalence*. As shown in Figure App.1, if you play a C on a piano keyboard, and then play the C an octave above it, you will notice that these two tones have a certain perceptual similarity. The same holds when you play a C followed by the C an octave below it—or any other C. Because of this strong perceptual similarity, tones that stand in octave relation are given the same name, such as C, C♯, D, D♯, and

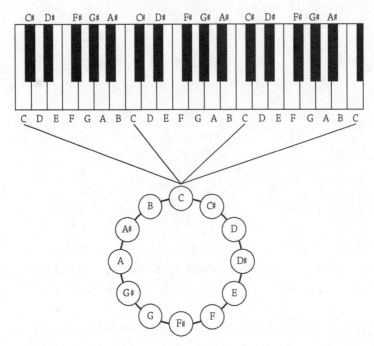

FIGURE APP.1. Illustration of a piano keyboard, showing the different notes. The Western musical scale is divided into octaves, with each octave consisting of 12 semitones—the notes C, C♯, D, D♯, E, F, F♯, G, G♯, A, A♯, B. Notes that stand in octave relationship are given the same name, and are described as being in the same pitch class.

so on, and music theorists describe tones that stand in octave relation as being in the same *pitch class*. As Figure App.1 depicts, we can think of pitch classes as forming a circle. Musical pitch therefore has two dimensions: the monotonic dimension of *pitch height* and the circular dimension of *pitch class*.

Tones that are related by octaves stand in the frequency ratio of 2:1. In the traditional musical scale, the octave is divided into 12 equal logarithmic steps, called *semitones*. So if you begin by playing C, then move one step up the keyboard to C♯, then one step further up to D, and so on, you are moving up in semitone steps. The same sequence of note names then continues along successive octaves. In terms of frequency, each successively higher tone is approximately 6 percent higher in frequency than the one below it.[2]

When two pairs of tones are separated by the same number of semitones, they are perceived as forming the same relationship, called a *musical interval*. So tones that are a semitone apart, such as C♯–D, or G–G♯, are perceived as forming the same interval; as are tones that are four semitones apart, such as C–E, F–A, and so on. Table App.1 shows the twelve musical intervals within an octave that correspond to the number of semitones formed by a tone pair.

Because of the perceptual similarity between pairs of tones that form the same interval, we can transfer a melody from one pitch range to another (this is termed *transposition*). Provided the intervals between successive tones remain the same, the identical melody is perceived, even though the tones themselves are different. So for example, the sequence of tones E–D–C–D–E–E–E, and the sequence B–A–G–A–B–B–B (with the notes in each sequence played in

**TABLE APP.1.**

Basic intervals in Western tonal music.

| Semitone Distance | Name of Interval |
| --- | --- |
| 0 | unison |
| 1 | minor second |
| 2 | major second |
| 3 | minor third |
| 4 | major third |
| 5 | perfect fourth |
| 6 | augmented fourth, diminished fourth, tritone |
| 7 | perfect fifth |
| 8 | minor sixth |
| 9 | major sixth |
| 10 | minor seventh |
| 11 | major seventh |
| 12 | octave |

Note: The octave is divided into 12 tones that are equally spaced logarithmically. A semitone is the distance between two adjacent tones (such as C and C♯).

the same octave), both produce the melody "Mary had a little lamb" (see Figure App.2 A & B). So just as visual shapes are perceived as being the same when they are moved from one location to another in the visual field, so a melody retains its essential musical shape when it's transposed from one pitch range to another. (The tritone paradox, explored in Chapter 5, provides a surprising exception to this rule.)

The syntax of a musical passage is hierarchical in nature, but this differs from the type of hierarchy that is invoked in language. A grammar in language, as it is represented in the brain, contains a lexicon of words and the concepts they signify, together with rules for combining the words so as to convey how the concepts are related. More specifically, a phrase structure grammar of a sentence can be portrayed as a tree structure. As an example, we can take the English sentence "The angry dog bit the man." This is composed of a noun phrase (NP) and a verb phrase (VP). In turn, the noun phrase ("The angry dog") is composed of a determiner ("the"), an adjective ("angry"), and a noun ("dog"). The verb phrase is composed of a verb ("bit") and another noun phrase, which is composed of a determiner ("the") and a noun ("man"). This produces the tree structure shown in Figure App.3.[3]

(a)

(b)

FIGURE APP.2. "Mary Had a Little Lamb" in the key of C (Figure 2a) and the key of G (Figure 2b).

Appendix

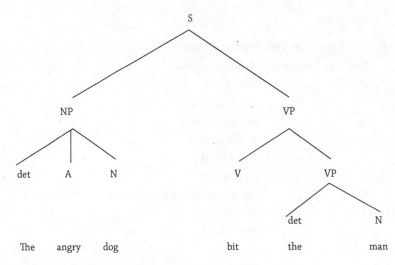

FIGURE APP.3. A sentence portrayed as a tree structure.

Because music doesn't have lexical meaning, it has no categories that correspond to nouns, adjectives, verbs, and so on. Instead, any note can in principle exist in any position in a musical hierarchy, depending on its relationships to the other notes in the passage.

Together with the mathematician John Feroe, I proposed a model of the way we represent pitch sequences in tonal music as hierarchies.[4] In essence, the model can be characterized as a hierarchical network, at each level of which structural units are represented as an organized set of elements. Elements that are present at any given level are elaborated by further elements so as to form structural units at the next-lower level, until the lowest level is reached. The model also assumes that Gestalt principles of perceptual organization, such as proximity and good continuation, contribute to organization at each hierarchical level.

According to the model, this hierarchical representation makes use of a basic layering of alphabets of notes, such as shown in Figure App.4. In classical Western tonal music, the lowest layer consists of the notes [C, C#, D, D#, E, F, F#, G, G#, A, A#, B, C,] known as the *chromatic scale*. This is the basic alphabet from which higher level alphabets are derived. The next-higher layer consists of a subset of the notes from the chromatic scale, and forms a *major or minor scale*.

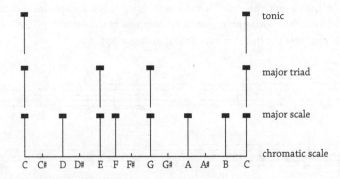

FIGURE APP.4. A hierarchy of embedded pitch alphabets.
From Deutsch & Feroe (1981).

So if the passage is in the key of C major, the second layer consists of the notes [C, D, E, F, G, A, B, C.] These notes then tend to be heard more prominently than the remaining notes of the chromatic scale, [C♯, D♯, F♯, G♯, A♯]. At the next-higher level, subsets of the scale, generally *triads*, are invoked. The triad beginning on the first note of the scale (called the *tonic*) is given most prominence. For example, the triad on the tonic of the C major scale—the notes [C, E, G, C] —tends to be heard particularly prominently. The highest level consists of the tonic alone—in this case it is the note C.

As an example of the model, we can consider the passage shown in Figure App.5(b). In principle, this could be represented in terms of steps that traverse the chromatic scale. A basic subsequence that consists of one step up this scale is presented four times in succession; the second presentation being four steps up from the first, the third being three steps up from the second, and the fourth being five steps up from the third. However, this analysis does not relate the various notes to each other in a musically meaningful way.

In contrast, a musical analysis of this passage would describe it as on two hierarchical levels, in which a higher level subsequence is elaborated by a lower level subsequence. The notes at the higher level, shown in Figure App.5(a), ascend stepwise through the C major triad (the notes C–E–G–C). At the lower level, shown in Figure App.5(b), each note is preceded by a note

FIGURE APP.5. A series of pitches at two hierarchical levels. (a) At the higher level, the notes move stepwise along the alphabet of the C major triad. (b) At the lower level, each note of the triad is preceded by a note a semitone lower (one step down the chromatic scale). (c) This pattern represented as a tree structure. From Deutsch & Feroe (1981).

| Beats | 1 | 2 | 3 | 4 | 1 | 2 | 3 | 4 | 1 | |
|---|---|---|---|---|---|---|---|---|---|---|
| | x | | | | x | | | | x | Level 1 |
| | x | | x | | x | | x | | x | Level 2 |
| | x | x | x | x | x | x | x | x | x | Level 3 |

FIGURE APP.6. A metrical hierarchy at three levels in 4/4 time, From Lerdahl & Jackendoff, 1983.

one step lower on the chromatic scale, so forming a two-note pattern. Figure App.5(c) shows this structure in diagram form.

As illustrated in this passage, the model has a number of advantages for perceptual and cognitive processing. First, the principle of proximity is invoked throughout: At the higher level the notes move in proximal steps along the C major triad [C–E–G–C]. At the lower level, each subsequence consists of one step up the chromatic scale [B–C] [D#–E] [F#–G] [B–C]. Also at the lower level the identical pattern is repeatedly presented, and this enables the listener to grasp the passage, and to remember it easily.[4-6]

ᴖ

So far, we have been considering the pitch relationships formed by notes in a passage. But rhythm is also an important characteristic of a musical phrase. If you tap out the rhythm of a familiar melody, such as "Happy Birthday," people are likely to recognize the passage based on its rhythm alone. Rhythm has two aspects: meter and grouping. *Meter* refers to the regular cycle of strong and weak beats that characterizes a passage. So in marches, for instance, the strong beat occurs on every other beat—"ONE, two, ONE, two"—or once every four beats— "ONE, two, three, four, ONE, two, three, four"—(think "Yankee Doodle"). In waltzes, a strong beat occurs once in every three beats—"ONE, two, three, ONE, two, three" (think "The Blue Danube"). Superimposed on meter is a rhythmic pattern that defines the *grouping* of notes.

Meter and grouping do not necessarily coincide. For example, in the hymn "Amazing Grace" a strong beat occurs once in every three beats, but the boundary between the two rhythmic phrases corresponding to "Amazing grace" and "how sweet the sound" cuts across the pattern of strong and weak beats defined by the meter. The music theorist Justin London's book *Hearing in Time*[7] provides a detailed exploration of the complexities involved in the perception of timing in music.

More specifically, in meter, the listener infers from the music a pattern of strong and weak beats, which mark off equal time spans. Beats are arranged in a *metrical hierarchy*. So if a beat is strong at one level, it's also a beat at the next higher level, and so on as the hierarchy is ascended. At all structural levels, the beats are equally spaced in time. Figure App.6 shows a *metrical grid*,[8] which represents the pattern of beats in 4/4 time at three structural levels. All four beats are represented at level 3. The first and third beats are heard as stronger than the second and fourth beats, and so are represented at level 2. The repeating first beat is heard most strongly, and so is represented at level 1. Listeners tend to focus primarily on beats at one or two intermediate levels of the metrical hierarchy, called the *tactus*.

ᴖ

Given that we organize musical passages as syntactic structures, when our perceptual system decides that a phrase is being sung rather than spoken, neural systems are invoked that produce and maintain the syntactic representation of the musical phrase. And since the words forming the sung phrase (the lyrics) have a different syntactic structure (that of speech), our hearing mechanism needs to process two different syntactic structures in order to comprehend the sung phrase in its entirety.

# NOTES

## INTRODUCTION

1. See Aristotle's discussions of illusions in *Parva Naturalia*; fourth century B.C.
2. Donald Macleod (personal communication), Andrew Oxenham (personal communication).
3. Deutsch, 1969.
4. Deutsch, 1972b.
5. Fitch, 2010, p. 6
6. For this and other historical accounts of scientific approaches to the study of sound and music, see Hunt, 1978; Cohen, 1984; Deutsch, 1984; Heller, 2013.
7. Boethius, *De Institutione Musica,* sixth century A.D., trans Calvin Bower, p. 57.

## CHAPTER 1

1. Strauss, 1949, p. 44.
2. Shakespeare, *The Tempest*, Act IV, Scene i, 148–150, 156–158.
3. Letter from Charles Villers to Georges Cuvier written in 1802. Cited in M. J. P. Flourens in *Phrenology Examined*, 1846, p. 102.
4. Oldfield, 1971.
5. Varney & Benton, 1975.
6. Luria, 1969.
7. Mebert & Michel, 1980.
8. Letter from M.C. Escher to Bruno Ernst on Oct 12, 1956. Cited in Ernst, 2018, p. 21.
9. The unusual performance practices of McCartney and Hendrix were described by Michael A. Levine (personal communication).
10. This story was relayed to me by the Hungarian violinist Janos Negesy.
11. Strauss, 1949, p. 44.

12. Oldfield, 1971.

13. Deutsch, 1970b.

14. For this experiment I used the Handedness Inventory in Oldfield, 1971.

15. Deutsch, 1978.

16. Deutsch, 1980a.


CHAPTER 2

1. Deutsch, 1974a, 1974b, 1975a.

2. Deutsch, 1983b.

3. Deutsch 1975a, 1981, 1983a; Deutsch & Roll, 1976. For studies concerning the neurological bases of the "what" and "where" components of the octave illusion, see Lamminmaki & Hari, 2000; Lamminmaki et al., 2012.

4. Treisman & Gelade, 1980.

5. Necker, 1832, p. 336.

6. Escher woodcut: *Regular Division of the Plane III,* 1957.

7. Wallach et al., 1949.

8. Franssen, 1960, 1962; see also Hartmann & Rakert, 1989.

9. Eric J. Heller described his discovery of this illusion in an email to me on May 21, 2018.

10. Deouell & Soroker, 2000.

11. Rauschecker & Tian, 2000.

12. Deutsch, 1974c, 1975a, 1975c, 1983a, 1987, 1995.

13. The passage at the beginning of the last movement of Tchaikovsky's Sixth Symphony ("*Pathétique*") shown in this module was filmed for *NOVA*'s episode "What Is Music?" produced by WGBH Boston for PBS and released in 1989. I am grateful to David Butler for pointing me to the interchange between Tchaikovsky and Nikisch.

14. Butler, 1979.

15. For a study of the neurological basis of the scale illusion, see Kuniki et al., 2013.

16. Sloboda, 1985.

17. Deutsch, 1995.

18. Deutsch, 2003.

19. Hall et al., 2000.

20. Deutsch, 1970a, 1972a.

21. Thompson et al., 2001.

22. Helmholtz, 1954.

23. Deutsch et al., 2007; Deutsch, 1995.

24. Mudd, 1963.

25. Deutsch, 1983a, 1985, 1987.

26. Machlis, 1977; Deutsch, 1987.

27. Robert Boynton, personal communication.


CHAPTER 3

1. The Rubin vase illusion was created by Edgar Rubin around 2015. Earlier versions appear in eighteenth-century French prints, which are composed of portraits in which the outlines of vases facing each other appear on either side.

2. Ellis, 1938.

3. Deutsch, 1975a.

4. Bregman, 1990.

5. Robert Gjerdingen (1994) has pointed to important analogies in the Gestalt principles of perceptual organization as applied to both vision and music.

6. Miller & Heise, 1950.

7. Bregman & Campbell, 1971.

8. Van Noorden, 1975.

9. Dowling, 1973; see also Bey & McAdams, 2003.

10. Bower & Springston, 1970.

11. Deutsch, 1980b.

12. For a detailed account of timbre perception in music, see McAdams, 2013.

13. Wessel, 1979.

14. Warren et al., 1969.

15. The Kanizsa triangle was first published by Gaetano Kanizsa, an Italian psychologist, in 1955.

16. Miller & Licklider, 1950.

17. Dannenbring, 1976.

18. Warren et al., 1972.

19. Sasaki, 1980.

20. Ball, 2010. p. 162.

21. Shakespeare, *Antony and Cleopatra*, Act 4, Scene 14.

CHAPTER 4

1. The phrase "Escher for the Ear" is taken from Yam, 1996, p. 14.

2. Hofstatadter, 1979.

3. *Ascending and Descending*. Lithograph by M. C. Escher, 1960. Escher's lithograph *Drawing Hands*, shown in Chapter 1, Figure 6, is another Strange Loop.

4. *Penrose Staircase*, derived from Penrose & Penrose, 1958.

5. Shepard, 1964.

6. Drobisch, 1855. Various helical representations of pitch have since been proposed, for example by Ruckmick, 1929; Bachem, 1950; Revesz, 1954; Pikler, 1966; Shepard, 1964, 1965, 1982.

7. The perceived pitch of a harmonic complex tone, when the harmonics are adjacent to each other, is heard as corresponding to the fundamental frequency; see the Appendix for further details. Since a Shepard tone consists only of harmonics that are related by octaves, we don't perceive its pitch as corresponding to a fundamental frequency, and its perceived height is ambiguous.

8. Risset, 1971.

9. Kim usually pairs playing the Shepard scale on the piano with humming and whistling two parts of "Frère Jacques" simultaneously! Scott Kim, personal communication.

10. Burns, 1981; Teranishi, 1982; Nakajima et al., 1988.

11. Braus, 1995.

12. *Los Angeles Times*, 2009, Feb 4. Here, sound designer Richard King explains his use of the Shepard scale in the film *The Dark Knight*.

13. Guerrasio, 2017, explains how Hans Zimmer's musical score, written for Christopher Nolan's film *Dunkirk*, employed effects based on the Shepard scale.

14. Benade, 1976.

15. Deutsch et al., 2008.

16. Deutsch, 2010.

CHAPTER 5

1. A tritone is the interval formed by tones that are separated by six semitones, or half an octave. To imagine the sound of a tritone, consider the song "Maria" from *West Side Story*. The interval between "Ma" and "ri" is a tritone.

2. The tones were created by the audio software experts F. Richard Moore and Mark Dolson. Each tone consisted of six frequency components that were separated by octaves, and were generated under a bell-shaped spectral envelope. For each tone pair, the position of the spectral envelope was identical for the two tones. For details, see Deutsch et al, 1987.

3. Deutsch et al., 1987.

4. Deutsch et al., 1990.

5. Deutsch, 1991.

6. Giangrande, 1998.

7. Dawe et al., 1998.

8. Ragozzine & Deutsch, 1994.

9. Deutsch, 2007.

10. Deutsch et al., 2004b. The template proposed to influence the pitch range of the speaking voice, together with perception of the tritone paradox, is one of pitch class, and not simply of pitch. In other words, it concerns the position of a tone (or the pitch of a vowel) within the octave, not its position along a continuum from low to high. When men and women sing together, they generally sing in octave relation. Also, men and women in a given linguistic community speak in pitch ranges that are roughly in octave relation, and so in the same pitch class range.

11. See Dolson, 1994 for a review. The hypothesis is also discussed in detail in Chapter 6, on absolute pitch.

12. Majewski et al., 1972.

13. Deutsch, Le, et al., 2009.

14. Fitch, 2010.

15. Kim, 1989, p. 29.

16. Deutsch, 1992.

CHAPTER 6

1. Deutsch, 1990, p. 21.

2. Profita & Bidder, 1988.

3. Geschwind & Fusillo, 1966.

4. Terhardt & Seewann, 1983.

5. Levitin, 1994. For a related study, see Halpern, 1989.

6. Schellenberg & Trehub, 2003.

7. Smith & Schmuckler, 2008.

8. Van Hedger et al., 2016.

9. Saffran & Griepentrog, 2001.

10. Rubenstein, 1973, p. 14.

11. Baharloo et al., 1998.

12. Brady, 1970.

13. Acquiring absolute pitch presents a special problem for children who are learning to play transposing instruments—that is, instruments for which music is notated at pitches that differ from those that actually sound. For example, when the note C occurs in a score written

for a B♭ clarinet, the instrument will sound a B♭. My friend and colleague Trevor Henthorn described how he first learned to play the B♭ trumpet at school in fifth grade. The music teacher began by playing the note B♭; then he named it "C" and drew a musical staff with a C below it. Following this, he played the note C and named it "D"; and so on. Later, when children who had been learning to play the flute came to join them, Trevor discovered that the "trumpet C" that he had been taught was not the same as the "flute C." It was only later that he learned formally about transposing instruments.

14. Deutsch et al., 2006; Deutsch, Dooley, et al., 2009.

15. Lennenberg, 1967.

16. Chechik et al., 1998.

17. Curtiss, 1977.

18. Lane, 1976.

19. Itard's documentation of his attempts to educate Victor formed the basis of François Truffaut's film *L'Enfant sauvage*.

20. Bates, 1992; Dennis & Whitaker, 1976; Duchowny et al., 1996; Vargha-Khadem et al., 1997; Woods, 1983.

21. Newport, 1990.

22. *Ibid.*

23. Pousada, 1994.

24. Deutsch et al., 2004a.

25. Glanz, 1999.

26. Deutsch et al., 2006.

27. Miyazaki et al., 2012.

28. Deutsch, Dooley, et al., 2009.

29. Gervain et al., 2013.

30. Schlaug et al., 1995.

31. Loui et al., 2010.

32. Deutsch & Dooley, 2013.

33. Miyazaki, 1990.

34. Deutsch et al., 2013.

35. Takeuchi & Hulse, 1991.

36. I am grateful to David Huron for compiling the frequency of occurrence of each pitch class in Barlow and Morgenstern's *Electronic Dictionary of Musical Themes*.

37. Lee & Lee, 2010.

38. Sergeant, 1969.

39. Hedger et al., 2013.

40. Deutsch et al., 2017. The algorithm we used followed from an earlier study involving subjects who did not possess absolute pitch (Deutsch, 1970a).

41. Dooley & Deutsch, 2010.

42. Dooley & Deutsch, 2011.

43. Vernon, 1977.

44. Monsaingeon, 2001, p. 140.

45. Braun & Chaloupka, 2005.

46. Hamilton et al., 2004.

47. Pascual-Leone et al., 2005.

48. Mottron et al., 2009; Heaton et al, 2008.

49. Rojas et al., 2002.

CHAPTER 7

1. Bolivar, Missouri *News*, October 11, 2008.

2. Oceanside APB news, April 12, 2008

3. A sound spectrum can be defined as the distribution of energy for a sound source, as a function of frequency.

4. Steven Pinker discusses Mondegreens and related mishearings in detail in his engaging book *The Language Instinct*. The last two mishearings were described by Michael A. Levine (personal communication).

5. Gombrich, 1986, pp. 77–78. Joe Banks discusses events surrounding Gombrich's memo in his book *Rorchach Audio* (2012).

6. Ben Burtt describes his use of a cheese casserole to create the impression of slithering snakes at https://www.youtube.com/watch?v=dYdFiYGLr-s.

7. Deutsch, 1995, 2003.

8. Carlson, 1996.

9. The Rorschach test was created by the Swiss psychologist Hermann Rorschach in the first part of the twentieth century. A set of cards, each of which contains a symmetrical ink blot, is presented to the subject, who reports what he or she sees. The subject's reports are interpreted as clues concerning his or her cognition, personality, and motivations.

10. Hernandez, 2018

11. Banks, 2012.

12. Grimby, 1993.

13. Shermer, 2009.

14. The Helix Nebula, also known as "The Eye of God" and formally the Helix, NGC 7293, is a large planetary nebula located in the constellation Aquarius. This image was taken using the Hubble Space Telescope.

15. Sung Ti, quoted in Gombrich, 1960, p. 188.

16. Leonardo da Vinci, 1956. Cited in Gombrich, 1960, p. 188.

17. Merkelbach & Van den Ven, 2001.

18. Warren, 1983.

19. The term *backward masking* is used in psychoacoustics to refer to an effect in which a brief, soft sound is followed immediately by another, louder sound, and perception of the first sound is suppressed perceptually.

20. Vokey & Read, 1985.

21. Remez et al., 1981.

22. Heller, 2013. Also visit the book's companion website, http://whyyouhearwhatyouhear.com.

23. McGurk & MacDonald, 1976.

24. Wright, 1969, p. 26.

25. Spooner lived from 1844 to 1930. Of famous spoonerisms, many were invented by others, but a few are believed to have been created by Spooner himself.

26. Baars et al., 1975.

27. Erard, 2007.


CHAPTER 8

1. Kellaris, 2003.

2. Williamson et al., 2012

3. Kraemer et al., 2005.

4. fMRI (functional magnetic resonance imaging) is a neuroimaging technique that enables researchers to see images of changing blood flow to brain regions, which result in changes in blood oxygen level in these regions. This device therefore detects changes in neural activity in brain structures. Another widely used brain mapping technique is MEG (magnetoencephalaography)—the recording of magnetic fields in the brain using very sensitive recording devices. These techniques are used both in basic research on perception and cognition, and also to localize pathological brain regions.

5. "Pipedown: The Campaign for Freedom From Piped Music." *Pipedown News*, July 16, 2017.

6. Barenboim, 2006.

7. "No Music Day? Enjoy the Silence . . ." *The Guardian*, November 21, 2007.

8. O'Malley, 1920, p. 242.

9. Berlin, 1916, p. 695.

10. Margulis, 2014.

11. Zajonc, 1968.

12. Winkelman et al., 2003.

13. Pereira et al., 2011.

14. Nunes et al., 2015.

15. Sacks, 2008.

16. Twain, 1876.

17. Burnt Toast Podcast, March 9, 2017.

18. Michael A. Levine (personal communication).

CHAPTER 9

1. The legend behind Giuseppe Tartini's "Devil's Trill Sonata." is illustrated in Figure 9.2: *Tartini's Dream*, by Louis-Léopold Boilly, Bibliothèque Nationale de France, 1824.

2. Esquirol, 1845.

3. Baillarger, 1846.

4. Griesinger, 1867.

5. Hughlings Jackson, 1884.

6. Charles Bonnet, 1760.

7. de Morsier, 1967.

8. Meadows & Munro, 1977, Gersztenkorn & Lee, 2015

9. Penfield & Perot, 1963.

10. The term *mashup* refers to a technique used in popular music whereby a recording is generated by combining and synchronizing tracks from different songs.

11. White, 1964, pp. 134–135.

12. Bleich & Moskowits, 2000.

13. Geiger, 2009; Shermer, 2011.

14. Simpson, 1988.

15. Sacks, 2012.

16. Lindbergh, 1953. See also Suedfeld & Mocellin, 1987.

17. Cellini, 1558/1956

18. Marco Polo, 2008 (original *c.* 1300).

19. Grimby, 1993.

20. Rosenhan, 1973.

21. Sidgwick et al., 1894.

22. Tien, 1991.

23. McCarthy-Jones, 2012.

24. McCarthy-Jones, 2012; Lurhmann, 2012.

25. Daverio, 1997.

26. Bartos, 1955, p. 149.

27. Monsaingeon, 2001, pp. 140, 142.

28. Budiansky, 2014

29. Personal communications from Michael A. Levine, Peter Burkholder, and Stephen Budiansky.

30. Budiansky, 2014, p. 31.

CHAPTER 10

1. Letter from Mussorgsky to Rimsky-Korsakov, July 1868. In Leyda & Bertensson, 1947.

2. For discussions concerning modular processing of speech and of music, see Liberman & Mattingly, 1989; Liberman & Whalen, 2000; Peretz & Morais, 1989; Peretz & Coltheart, 2003; Zatorre et al., 2002.

3. For discussions of the concept of modularity in general, see Fodor, 1983, 2001.

4. Spencer, 1857.

5. Letter from Mussorgsky to Rimsky-Korsakov, July 1868. In Leyda & Bertensson, 1947.

6. Letter from Mussorgsky to Shestakova, July 1868. Op. cit.

7. Jonathan & Secora, 2006.

8. Luria et al., 1965.

9. Basso & Capitani, 1985.

10. Mazzuchi et al., 1982.

11. Piccarilli et al., 2000.

12. Darwin, 1958, pp. 20, 21.

13. Allen, 1878.

14. Peretz et al., 2007.

15. Peretz et al., 2002.

16. Loui et al., 2009.

17. Rex's amazing musical abilities are displayed in TV interviews with Lesley Stahl on CBS, 60 *Minutes*, which aired in 2003, 2005, and 2008.

18. See note 4, Chapter 8.

19. Koelsch et al., 2002.

20. Patel, 2008.

21. Norman-Haignere et al., 2015.

22. Deutsch, 1970b.

23. Deutsch, 1975b; 2013c.

24. Semal & Demany, 1991; Demany & Semal, 2008.

25. Starr & Pitt, 1997; Mercer & McKeown, 2010.

26. Pinker, 1997, p. 30.

27. Simon, 1962.

28. Fitch, 2010, pp. 17–18.

29. Deutsch et al., 2011.

30. Vanden Bosch der Nederlanden et al. (2015b) found that when subjects were asked to discriminate between the initial and final repetitions, they were better able to detect pitch

changes that violated the pitch structure of Western tonal music than with changes that accorded with this pitch structure. This finding is as expected if the subjects were hearing the phrase as music rather than speech.

31. Michael A. Levine, personal communication.

32. Shakespeare, *Julius Caesar*, Act III, Scene ii.

33. Speech delivered by Winston Churchill to the House of Commons on June 4, 1940, early in World War II.

34. Speech delivered by Martin Luther King, Jr. on the steps of the State Capitol in Montgomery, Alabama, on March 25, 1965.

35. Video of chanting farm workers.

36. Speech delivered by Barack Obama on January 8, 2008, in Nashua, New Hampshire, during his presidential campaign.

37. Robert Frost, "Stopping by Woods on a Snowy Evening," from Frost, 1969.

38. See Appendix for a description of the harmonic series.

39. Vanden Bosch der Nederlanden et al. (2015a).

40. Falk et al., 2014.

41. Margulis et al., 2015.

42. Jaisin et al., 2016.

43. Tierney et al., 2013. The term "pitch salience" indicates how prominently pitch is perceived as a feature of sound.

44. See Hymers et al. (2015) for further work on the neurological basis of the speech-to-song illusion.

CHAPTER 11

1. Spencer, 1857.

2. Steele, 1779.

3. Darwin, 1879.

4. Jespersen, 1922.

5. Wray, 1998.

6. For a detailed discussion of the relationship between linguistic prosody and musical structure, see Heffner & Slevc, 2015.

7. Bitterns are wading birds in the heron family. They are secretive and difficult to see, but they make remarkably loud, low, booming sounds, rather like a foghorn.

8. Shaw, 1913, Act 1.

9. Juslin & Laukka, 2003.

10. Querleu et al., 1988.

11. DeCaspar et al., 1994.

12. DeCaspar & Fifer, 1980.

13. Spence & Freeman, 1996.

14. Moon et al., 1993.

15. Mampe et al., 2009.

16. Fernald, 1993.

17. Thiessen et al., 2005.

18. Moreno et al., 2009.

19. Thompson et al., 2004.

20. Marques et al., 2007.

21. Lima & Castro, 2011.

22. Thompson et al., 2012.

23. Wong et al., 2007.

24. Schellenberg, 2015.

25. Giuliano et al., 2011; Pfordresher & Brown, 2009; Creel et al., 2018.

26. Krishnan et al., 2005.

27. Bidelman et al., 2013.

28. Marie et al., 2012.

29. Patel & Daniele, 2003.

30. Huron & Ollen (2003) used a greatly expanded sample of themes, and replicated Patel and Daniele's finding that note-to-note contrast in duration was higher in English than in French music.

31. Iversen et al., 2008.

32. See Chapter 6 for a description of the different Mandarin tones.

33. https://chinesepod.com/blog/2014/10/25/how-to-read-a-chinese-poem-with-only-one-sound/

34. The origin of this sentence is unclear, but the linguist Robert Berwick (personal communication) has told me that he had heard this sentence in 1972 as an eleven year old. I am also grateful to Steven Pinker for a discussion of this issue, and his book *The Language Instinct* points to the similar sentence "Buffalo buffalo Buffalo buffalo buffalo buffalo Buffalo buffalo" (pp. 209–210).

35. Michael A. Levine, personal communication.

36. Darwin, 1871, p. 55.

37. Mithen, 2006.

38. Brown, 2000.

APPENDIX

1. For a detailed discussion of the harmonic series, see Oxenham, 2013; Hartmann, 2005; Sundberg, 2013; Rossing et al., 2002; and Heller, 2013, including its companion website http://whyyouhearwhatyouhear.com.

2. Each note that corresponds to a black key on the piano is named either a sharp (♯) or a flat (♭), indicating that it is a semitone higher than the white note below it, or a semitone lower than the white note above it. So the black note on the piano just above C can be labeled either C♯ or D♭.

3. Pinker, 1994. See also Pinker, 1997.

4. Deutsch & Feroe, 1981, Deutsch, 1999.

5. The music theorist Fred Lerdahl (2001) has proposed a "basic space" that is a rendering of Deutsch and Feroe's hierarchy of alphabets. Lerdahl's pitch space is elaborated to include chords and keys.

6. For further discussions of tonal structure in music, see Schenker, 1956; Meyer, 1956; Deutsch, 2013b; Gjerdingen, 1988; Krumhansl, 1990; Huron, 2016; Narmour, 1990, 1992; Temperley, 2007; Thompson, 2013; Thomson, 1999.

7. London, 2004; see also Honing, 2013.

8. Lerdahl & Jackendoff, 1983.

## REFERENCES

Allen, C. (1878). Note-deafness, *Mind*, 3, 157–167.

Aristotle. (1955). *Parva naturalia* (W. D. Ross, Ed.). Oxford: Clarendon Press.

Baars, B. J., Motley, M. T., & MacKay, D. G. (1975). Output editing for lexical status in artificially elicited slips of the tongue. *Journal of Verbal Learning and Verbal Behavior*, 14, 382–391.

Bachem, A. (1950). Tone height and tone chroma as two different pitch qualities. *Acta Psychologica*, 7, 80–88.

Baharloo, S., Johnston, P. A., Service, S. K., Gitschier, J., & Freimer, N. B. (1998). Absolute pitch: An approach for identification of genetic and nongenetic components, *American Journal of Human Genetics*, 62, 224–231.

Baillarger, M. J. (1846). Des hallucinations. *Memoires l'Academie Royale de Medicine*, 12, 273–475.

Ball, P. (2010). *The music instinct: How music works and why we can't do without it*. New York: Oxford University Press.

Banks, J. (2012). *Rorschach audio: Art & illusion for sound*. London: Strange Attractor Press.

Barenboim, D. (2006). *In the beginning was sound*. Reith Lecture.

Bartos, F. (1955). *Bedřich Smetana: Letters and reminiscences* (D. Rusbridge, Trans.). Prague: Artia.

Basso, A., & Capitani, E. (1985). Spared musical abilities in a conductor with global aphasia and ideomotor apraxia. *Journal of Neurology, Neurosurgery and Psychiatry*, 48, 407–412.

Bates, E. (1992). Language development. *Current Opinion in Neurobiology*, 2, 180–185.

Benade, A. H. (1976). *Fundamentals of musical acoustics*. New York: Oxford University Press.

Bey, C., & McAdams, S. (2003). Postrecognition of interleaved melodies as an indirect measure of auditory stream formation. *Journal of Experimental Psychology: Human Perception and Performance*, 29, 267–279.

Berlin, I. (1916). Love-interest as a commodity. *Green Book Magazine*, 15, 695–698.

Bidelman, G. M., Hutka, S., & Moreno, S. (2013). Tone language speakers and musicians share enhanced perceptual and cognitive abilities for musical pitch: Evidence for bidirectionality between the domains of language and music. *PLoS ONE, 8*, e60676.

Bleich, A., & Moskowits, L. (2000). Post traumatic stress disorder with psychotic features. *Croatian Medical Journal, 41*, 442–445.

Boethius, A. M. S. (ca. 500–507). *De institutione musica.* (Calvin Bower, Trans., 1966).

Bonnet, C. (1760). *Essai analytique sur les faculties de l'ame.* Copenhagen: Philbert.

Bower, G. H., & Springston, (1970). Pauses as recoding points in letter series. *Journal of Experimental Psychology, 83*, 421–430.

Brady, P. T. (1970). Fixed scale mechanism of absolute pitch. *Journal of the Acoustical Society of America, 48*, 883–887.

Braun, M., & Chaloupka, V. (2005). Carbamazepine induced pitch shift and octave space representation. *Hearing Research, 210*, 85–92.

Braus, I. (1995). Retracing one's steps: An overview of pitch circularity and Shepard tones in European music, 1550–1990. *Music Perception, 12*, 323–351.

Bregman, A. S. (1990). *Auditory scene analysis: The perceptual organization of sound.* Cambridge, MA: MIT Press.

Bregman, A. S., & Campbell, J. (1971). Primary auditory stream segregation and perception of order in rapid sequences of tones. *Journal of Experimental Psychology, 89*, 244–249.

Brown, S. (2000). The "Musilanguage" model of music evolution. In N. L. Wallin, B. Merker, & S. Brown (Eds.), *The origins of music* (pp. 271–300). Cambridge, MA: MIT Press.

Budiansky, S. (2014). *Mad music: Charles Ives, the nostalgic rebel.* Lebanon, NH: University Press of New England.

Burns, E. (1981). Circularity in relative pitch judgments for inharmonic complex tones: The Shepard demonstration revisited, again. *Perception & Psychophysics, 30*, 467–472.

Butler, D. (1979). A further study of melodic channeling. *Perception & Psychophysics, 25*, 264–268.

Carlson, S. (1996, December). Dissecting the brain with sound. *Scientific American*, 112–115.

Cellini, B. (1956). *Autobiography* (G. Bull, Trans.). London: Penguin Books. (Original work published ca. 1558)

Chechik, G., Meilijson, I., & Ruppin, E. (1998). Synaptic pruning in development: A computational account. *Neural Computation, 10*, 1759–1777.

Chen Yung-chi, quoted from H. A. Giles, *An introduction to the history of Chinese pictorial art* (Shanghai and Leiden, 1905), p. 100.

Creel, S. C., Weng, M., Fu, G., Heyman, G. D., & Lee, K. (2018). Speaking a tone language enhances musical pitch perception in 3–5-year-olds. *Developmental Science, 21*, e12503.

Curtiss, S. (1977). *Genie: A psycholinguistic study of a modern day "wild child."* New York: Academic Press.

Dannenbring, G. L. (1976). Perceived auditory continuity with alternately rising and falling frequency transitions. *Canadian Journal of Psychology, 30*, 99–114.

Darwin, C. R. (1958). *The Autobiography of Charles Darwin, and Selected Letters* (F. Darwin, Ed.). New York: Dover. (Originally published in 1892.)

Darwin, C. R. (2004). *The Descent of Man, and Selection in Relation to Sex.* 2nd ed. London: Penguin Books. (Originally published 1879 by J. Murray, London.)

Daverio, J. (1997). *Robert Schumann: Herald of a "new poetic age."* New York: Oxford University Press.

Dawe, L. A., Platt, J. R., & Welsh, E. (1998). Spectral-motion aftereffects and the tritone paradox among Canadian subjects. *Perception & Psychophysics, 60*, 209–220.

DeCaspar, A. J., & Fifer, W. P. (1980). Of human bonding: Newborns prefer their mothers' voice. *Science*, 208, 1174–1176.

DeCaspar, A. J., Lecanuet, J.-P., Busnel, M.-C., Granier-Deferre, C., & Maugeais, R. (1994). Fetal reactions to recurrent maternal speech. *Infant Behavior and Development*, 17, 159–164.

Deouell, L. Y., & Soroker, N. (2000). What is extinguished in auditory extinction? *NeuroReport*, 2000, 11, 3059–3062.

Demany, L., & Semal, C. (2008). The role of memory in auditory perception. In W. A. Yost, A. N. Popper, & R. R. Fay (Eds.), *Auditory perception of sound sources* (pp. 77–113). New York: Springer.

Dennis, M., & Whitaker, H. A. (1976). Language acquisition following hemidecortication: linguistic superiority of the left over the right hemisphere. *Brain and Language*, 3, 404–433.

de Morsier, G. (1967). Le syndrome de Charles Bonnet: hallucinations visuelles des vieillards sans deficience mentale [Charles Bonnet syndrome: visual hallucinations of the elderly without mental impairment]. *Ann Med Psychol* (in French), 125, 677–701.

Deutsch, D. (1969). Music recognition. *Psychological Review*, 76, 300–309.

Deutsch, D. (1970a). Dislocation of tones in a musical sequence: A memory illusion. *Nature*, 226, 286.

Deutsch, D. (1970b). Tones and numbers: Specificity of interference in immediate memory. *Science*, 168, 1604–1605

Deutsch, D. (1972a). Effect of repetition of standard and comparison tones on recognition memory for pitch. *Journal of Experimental Psychology*, 93, 156–162.

Deutsch, D. (1972b) Octave generalization and tune recognition. *Perception & Psychophysics*, 11, 411–412.

Deutsch, D. (1974a). An auditory illusion. *Journal of the Acoustical Society of America*, 55, s18–s19.

Deutsch, D. (1974b). An auditory illusion. *Nature*, 251, 307–309.

Deutsch, D. (1974c). An illusion with musical scales. *Journal of the Acoustical Society of America*, 56, s25.

Deutsch, D. (1975a). Musical illusions. *Scientific American*, 233, 92–104.

Deutsch, D. (1975b). The organization of short term memory for a single acoustic attribute. In D. Deutsch & J. A. Deutsch (Eds.). *Short term memory* (pp. 107–151). New York: Academic Press.

Deutsch, D. (1975c). Two-channel listening to musical scales. *Journal of the Acoustical Society of America*, 57, 1156–1160.

Deutsch, D. (1978). Pitch memory: An advantage for the lefthanded. *Science*, 199, 559–560.

Deutsch, D. (1980a). Handedness and memory for tonal pitch. In J. Herron (Ed.), *Neuropsychology of lefthandedness* (pp. 263–271). New York: Academic Press.

Deutsch, D. (1980b). The processing of structured and unstructured tonal sequences. *Perception & Psychophysics*, 28, 381–389.

Deutsch, D. (1980c). Music perception. *The Musical Quarterly*, 66, 165–179.

Deutsch, D. (1981). The octave illusion and auditory perceptual integration. In J. V. Tobias & E. D. Schubert (Eds.), *Hearing research and theory* (Vol. I, pp. 99–142). New York: Academic Press.

Deutsch, D. (1983a). Auditory illusions, handedness, and the spatial environment. *Journal of the Audio Engineering Society*, 31, 607–618.

Deutsch, D. (1983b). The octave illusion in relation to handedness and familial handedness background. *Neuropsychologia*, 21, 289–293.

Deutsch, D. (1984). Psychology and music. In M.H. Bornstein (Ed.). *Psychology and its allied disciplines*, 1984, 155–194, Hillsdale: Erlbaum.

Deutsch, D. (1985). Dichotic listening to melodic patterns and its relationship to hemispheric specialization of function. *Music Perception, 3*, 127–154.

Deutsch, D. (1986a). An auditory paradox. *Journal of the Acoustical Society of America, 80*, s93.

Deutsch, D. (1986b). A musical paradox. *Music Perception, 3*, 275–280.

Deutsch, D. (1987). Illusions for stereo headphones. *Audio Magazine*, 36–48.

Deutsch, D. (1991). The tritone paradox: An influence of language on music perception. *Music Perception, 8*, 335–347.

Deutsch, D. (1992). Paradoxes of musical pitch. *Scientific American, 267*, 88–95.

Deutsch, D. (1995). *Musical illusions and paradoxes* [Compact disc and booklet]. La Jolla, CA: Philomel Records.

Deutsch, D. (1996). The perception of auditory patterns. In W. Prinz and B. Bridgeman (Eds.), *Handbook of perception and action* (Vol. 1, pp. 253–296). Orlando, FL: Academic Press.

Deutsch, D. (1997). The tritone paradox: A link between music and speech. *Current Directions in Psychological Science, 6*, 174–180.

Deutsch, D. (Ed.). (1999). The processing of pitch combinations. In D. Deutsch (Ed.), *The Psychology of music* (2nd ed. pp. 349–412). San Diego, CA: Elsevier.

Deutsch, D. (2003). *Phantom words, and other curiosities* (Audio CD and booklet). La Jolla, CA: Philomel Records.

Deutsch, D. (2007). Mothers and their offspring perceive the tritone paradox in closely similar ways. *Archives of Acoustics, 32*, 3–14.

Deutsch, D. (2010, July 8–15). The paradox of pitch circularity. *Acoustics Today*, 8–14.

Deutsch, D. (2013a). Grouping mechanisms in music. In D. Deutsch (Ed.), *The psychology of music* (3rd ed., pp. 183–248). San Diego, CA: Elsevier.

Deutsch, D. (Ed.). (2013b). *The psychology of music* (3rd ed.). San Diego, CA: Elsevier.

Deutsch, D. (2013c). The processing of pitch combinations. In D. Deutsch (Ed.) *The psychology of music* (3rd ed., pp. 249–324). San Diego, CA: Elsevier.

Deutsch, D., & Dooley, K. (2013). Absolute pitch is associated with a large auditory digit span: A clue to its genesis. *Journal of the Acoustical Society of America, 133*, 1859–1861,

Deutsch, D., Dooley, K., & Henthorn, T. (2008). Pitch circularity from tones comprising full harmonic series. *Journal of the Acoustical Society of America, 124*, 589–597.

Deutsch, D., Dooley, K., Henthorn, T., & Head, B. (2009). Absolute pitch among students in an American music conservatory: Association with tone language fluency. *Journal of the Acoustical Society of America, 125*, 2398–2403.

Deutsch, D., Edelstein, M., & Henthorn, T. (2017). Absolute pitch is disrupted by an auditory illusion. *Journal of the Acoustical Society of America, 141*, 3800.

Deutsch, D., & Feroe, J. (1981). The internal representation of pitch sequences in tonal music. *Psychological Review, 88*, 503–522.

Deutsch, D., Henthorn, T., Marvin, E., & Xu, H.-S. (2006). Absolute pitch among American and Chinese conservatory students: Prevalence differences, and evidence for a speech-related critical period. *Journal of the Acoustical Society of America, 119*, 719–722.

Deutsch, D., Hamaoui, K., & Henthorn, T. (2007). The glissando illusion and handedness. *Neuropsychologia, 45*, 2981–2988.

Deutsch, D., Henthorn, T., & Dolson, M. (2004a). Absolute pitch, speech, and tone language: Some experiments and a proposed framework. *Music Perception, 21*, 339–356.

Deutsch, D., Henthorn, T., & Dolson, M. (2004b). Speech patterns heard early in life influence later perception of the tritone paradox. *Music Perception, 21*, 357–372.

Deutsch, D., Henthorn, T., & Lapidis, R. (2011). Illusory transformation from speech to song. *Journal of the Acoustical Society of America, 129,* 2245–2252.

Deutsch, D., Kuyper, W. L., & Fisher, Y. (1987). The tritone paradox: Its presence and form of distribution in a general population. *Music Perception, 5,* 79–92.

Deutsch, D., Li, X., & Shen, J. (2013). Absolute pitch among students at the Shanghai Conservatory of Music: A large-scale direct-test study. *Journal of the Acoustical Society of America, 134,* 3853–3859.

Deutsch, D., Le, J., Shen, J., & Henthorn, T. (2009). The pitch levels of female speech in two Chinese villages. *Journal of the Acoustical Society Express Letters, 125,* EL208–EL213.

Deutsch, D., North, T. & Ray, L. (1990). The tritone paradox: Correlate with the listener's vocal range for speech. *Music Perception, 7,* 371–384.

Deutsch, D. & Roll, P. L. (1976). Separate "what" and "where" mechanisms in processing a dichotic tonal sequence. *Journal of Experimental Psychology: Human Perception and Performance, 2,* 23–29.

Deutsch, E. O., (1990). *Mozart: A documentary biography* (3rd. ed.). New York: Simon & Schuster.

Dolson, M. (1994). The pitch of speech as a function of linguistic community. *Music Perception, 11,* 321–331.

Dooley, K., & Deutsch, D. (2010). Absolute pitch correlates with high performance on musical dictation. *Journal of the Acoustical Society of America, 128,* 890–893.

Dooley, K., & Deutsch, D. (2011). Absolute pitch correlates with high performance on interval naming tasks. *Journal of the Acoustical Society of America, 130,* 4097–4104.

Dowling, W. J. (1973). The perception of interleaved melodies. *Cognitive Psychology, 5,* 322–337.

Dowling, W. J., & Harwood, D. L. (1986). *Music cognition.* Orlando, FL: Academic Press.

Drobisch, M. W. (1855). Uber musikalische Tonbestimmung und Temperatur. In *Abhandlungen der Königlich sachsischen Gesellschaft der Wissenschaften zu Leipzig, 4,* 3–121. Leipzig: Hirzel.

Duchowny, M., Jayakar, P., Harvey, A. S., Resnick, T., Alvarez, L., Dean, P., & Levin, B. (1996). Language cortex representation: effects of developmental versus acquired pathology. *Annals of Neurology, 40,* 31–38.

Ellis, W. D. (1938). *A source book of Gestalt psychology.* London: Routledge and Kegan Paul.

Erard, M. (2007). *Um . . . Slips, stumbles, and verbal blunders, and what they mean.* New York: Random House.

Ernst, B. (1976). *The magic mirror of M. C. Escher.* New York: Random House.

Esquirol, E. (1845). *Mental maladies: A treatise on insanity* (E. K. Hunt, Trans.). Philadelphia: Lea & Blanchard. (Original French edition 1838)

Falk, S., Rathke, T., & Dalla Bella, S. (2014). When speech sounds like music. *Journal of Experimental Psychology, Human Perception and Performance, 40,* 1491–1506.

Fernald, A. (1993). Approval and disapproval: Infant responsiveness to vocal affect in familiar and unfamiliar languages. *Child Development, 64,* 657–674.

Fitch, W. T. (2010). *The evolution of language.* Cambridge, UK: Cambridge University Press.

Flourens, M. J. P. (1846). *Phrenology Examined* (Trans. C. de L. Meigs). Philadelphia, PA: Hogan & Thompson.

Fodor, J. A. (1983). *The modularity of mind: An essay on faculty psychology.* Cambridge, MA: MIT Press.

Fodor, J. A. (2001). *The mind doesn't work that way: The scope and limits of computational psychology.* Cambridge, MA: MIT Press.

Franssen, N. V. (1960). *Some considerations on the mechanism of directional hearing.* (Doctoral dissertation). Techniche Hogeschool, Delft, the Netherlands.

Franssen, N. V. (1962). *Stereophony.* Eindhoven, the Netherlands: Philips Technical Library. (English Translation 1964)

Frost, R. (1969). *The poetry of Robert Frost* (E. C. Lathem, Ed.). New York: Henry Holt.

Geiger, J. (2009). *The third man factor: The secret of survival in extreme environments.* New York: Penguin.

Gersztenkorn, D., & Lee, A. G. (2015). Palinopsia revamped: A systematic review of the literature. *Survey of Ophthalmology, 60,* 1–35.

Gervain, J., Vines, B. W., Chen, L. M., Seo, R. J., Hensch, T. K., Werker, J. F., & Young, A. H. (2013). Valproate reopens critical-period learning of absolute pitch, *Frontiers in Systems Neuroscience, 7,* 102.

Geschwind, N., & Fusillo, M. (1966). Color-naming defects in association with alexia. *Archives of Neurology, 15,* 137–146.

Giangrande, J. (1998). The tritone paradox: Effects of pitch class and position of the spectral envelope. *Music Perception, 15,* 253–264.

Giles, H. A. (1905). *An introduction to the history of Chinese pictorial art.* London: Bernard Quaritch.

Giuliano, R. J., Pfordresher, P. Q., Stanley, E. M., Narayana, S., & Wicha, N. Y. (2011). Native experience with a tone language enhances pitch discrimination and the timing of neural responses to pitch change. *Frontiers in Psychology, 2,* 146.

Gjerdingen, R. O. (1988). *A classic turn of phrase: Music and the psychology of convention.* Philadelphia: University of Pennsylvania Press.

Gjerdingen, R. O. (1994). Apparent motion in music? *Music Perception 11,* 335–370.

Glanz, J. (1999, November 5). Study links perfect pitch to tonal language. *New York Times.* Retrieved from https://www.nytimes.com/1999/11/05/us/study-links-perfect-pitch-to-tonal-language.html

Gombrich, E. H. (1986). Some axioms, musings, and hints on hearing. In O. Renier & V. Rubenstein, *Assigned to listen: The Evesham experience, 1939–43* (pp. 77–78). London: BBC External Services.

Gombrich, E. H. (2000). *Art and illusion: A study in the psychology of pictorial representation.* Princeton, NJ: Princeton University Press, eleventh printing.

Griesinger, W. (1867). *Mental pathology and therapeutics* (C. L. Robertson & J. Rutherford, Trans.) London: New Sydenham Society.

Grimby, A. (1993). Bereavement among elderly people: Grief reactions, post-bereavement hallucinations and quality of life. *Acta Psychiatrica Scandinavica, 87,* 72–80.

Guerrasio, J. (2017, July 24). Christopher Nolan explains the "audio illusion" that created the unique music in "Dunkirk." *Entertainment—Business Insider.* Retrieved from https://www.businessinsider.com/dunkirk-music-christopher-nolan-hans-zimmer-2017-7

Halpern, A. R. (1989). Memory for the absolute pitch of familiar songs. *Memory and Cognition, 17,* 572–581.

Hartmann, W. M. (2005). *Signals, sound, and sensation.* New York: Springer.

Hartmann, W. M., & Rakert, B. (1989). Localization of sound in rooms IV: The Franssen effect. *Journal of the Acoustical Society of America, 86,* 1366–1373.

Hall, M. D., Pastore, R. E., Acker, B. E., & Huang, W. (2000). Evidence for auditory feature integration with spatially distributed items. *Perception & Psychophysics, 62,* 1243–1257.

Hamilton, R. H., Pascual-Leone, A., & Schlaug, G. (2004). Absolute pitch in blind musicians. *NeuroReport*, 15, 803–806.

Hartmann, W. M. (2005). Signals, sound, and sensation. New York: Springer.

Heaton, P., Davis, R. E., & Happe, F. G. (2008). Research note: Exceptional absolute pitch perception for spoken words in an able adult with autism. *Neuropsychologia* 46, 2095–2098.

Hedger, S. C. Van, Heald, S. L. M., & Nusbaum, H. C. (2013). Absolute pitch may not be so absolute. *Psychological Science*, 24, 1496–1502.

Hedger, S. C., Van, Heald, S. L. M. & Nusbaum, H. C. (2016). What the [bleep]? Enhanced pitch memory for a 1000 Hz sine tone. Cognition, 154, 139–150.

Heffner, C. C., & Slevc, L. R. (2015). Prosodic structure as a parallel to musical structure. *Frontiers in Psychology*, 6, 1962.

Heller, E. J. (2013). *Why you hear what you hear: An experiential approach to sound, music, and psychoacoustics*. Princeton, NJ and Oxford, UK: Princeton University Press.

Helmholtz, H. von. (1954). *On the sensations of tone as a physiological basis for the theory of music* (2nd English ed.). New York: Dover. (Original work published 1877)

Hernandez, D. (2018, May 17). Yanny or Laurel? Your brain hears what it wants to. *Wall Street Journal*. Retrieved from https://www.wsj.com/articles/yanny-or-laurel-your-brain-hears-what-it-wants-to-1526581025

Hofstadter, D. R. (1979). *Gödel, Escher, Bach: An eternal golden braid*. New York: Basic Books.

Honing, H. (2013). Structure and interpretation of rhythm in music. In D. Deutsch (Ed.), *The psychology of music* (3rd ed., pp. 369–404). San Diego, CA: Elsevier.

Hunt, F. V. (1978) *Origins in acoustics: The science of sound from antiquity to the age of Newton*. London: Yale University Press, Ltd.

Huron, D., & Ollen, J. (2003). Agogic contrast in French and English themes: Further support for Patel and Daniele. (2003). *Music Perception*, 21, 267–271.

Hymers, M., Prendergast, G., Liu, C., Schulze, A., Young, M. L., Wastling, S. J., . . . Millman, R. E. (2015). Neural mechanisms underlying song and speech perception can be differentiated using an illusory percept. *NeuroImage*, 108, 225–233.

Iversen, J. R., Patel, A. D., & Ohgushi, K. (2008). Perception of rhythmic grouping depends on auditory experience. *The Journal of the Acoustical Society of America*, 124, 2263–2271.

Jackson, J. H. (1884). The Croonian Lectures on the evolution and dissolution of the nervous system. *British Medical Journal*, 1, 591–593.

Jaisin, K., Suphanchaimat, R., Candia, M. A. F., & Warren, J. D. (2016). The speech-to-song illusion is reduced in speakers of tonal (vs. non-tonal) languages. *Frontiers of Psychology*, 7, 662.

Jespersen, O. (1922). *Language: Its nature, development and origin*. London: Allen & Unwin Ltd.

Jonathan, G. & Secora, P. (2006). Eavesdropping with a master: Leos Janacek and the music of speech. *Empirical Musicology Review*, 1, 131–165.

Juslin, P.N. & Laukka, P. (2003). Communication of emotions in vocal expression and music performance: different channels, same code? *Psychological Bulletin*, 129, 770–814.

Kanizsa, G. (1955). Margini quasi-percettivi in campi con stimolazione omogenea. *Revista di Psicologia*, 49, 7–30.

Kellaris, J. (2003, February 22). *Dissecting earworms: Further evidence on the "song-stuck-in-your-head phenomenon."* Paper presented to the Society for Consumer Psychology, New Orleans, LA.

Kim, S. (1989). *Inversions*. New York: W. H. Freeman.

King, R. (2009, February 4) *"The Dark Knight"* sound effects. *Los Angeles Times*.

Koelsch, S., Gunter, T. C., Cramon, D. Y., Zysset, S., Lohmann, G., & Friederici, A. D. (2002). Bach speaks: A cortical "language-network" serves the processing of music. *NeuroImage, 17,* 956–966.

Kraemer, D. J. M., Macrae, C. N., Green, A. E., & Kelley, W. M. (2005). Sound of silence activates auditory cortex. *Nature, 434,* 158.

Krishnan, A., Xu, Y., Gandour, J., & Cariani, P. (2005). Encoding of pitch in the human brainstem is sensitive to language experience. *Cognitive Brain Research, 25,* 161–168.

Krumhansl, C. L. (1990). *Cognitive foundations of musical pitch.* New York: Oxford University Press.

Kuniki, S., Yokosawa, K., & Takahashi, M. (2013). Neural representation of the scale illusion: Magnetoencephalographic study on the auditory illusion induced by distinctive tone sequences in the two ears. *PLoS ONE* 8(9): e7599.

Lamminmaki, S., & Hari, R. (2000). Auditory cortex activation associated with the octave illusion. *Neuroreport, 11,* 1469–1472.

Lamminmaki, S., Mandel, A., Parkkonen, L., & Hari, R. (2012). Binaural interaction and the octave illusion. *Journal of the Acoustical Society of America, 132,* 1747–1753.

Lane, H. (1976). *The wild boy of Aveyron,* Cambridge, MA: Harvard University Press.

Lee, C.-Y., & Lee, Y.-F. (2010). Perception of musical pitch and lexical tones by Mandarin speaking musicians. *Journal of the Acoustical Society of America, 127,* 481–490.

Lennenberg, E. H. (1967) *Biological foundations of language.* New York: Wiley.

Leonardo da Vinci (1956). *Treatise on Painting* (A. P. McMahon, Ed.). Princeton, NJ: Princeton University Press.

Lerdahl, F. (2001). *Tonal pitch space.* Oxford, UK: Oxford University Press.

Lerdahl, F., & Jackendoff, R. (1983). *A generative theory of tonal music.* Cambridge, MA: MIT Press.

Leyda, J. & Bertensson, S. (1947) The Musorgsky reader: A life of Modeste Petrovich Musorgsky in letters and documents (Eds. & Trans.) New York: Norton.

Levitin, D. J. (1994). Absolute memory for musical pitch: Evidence from the production of learned melodies. *Perception & Psychophysics, 56,* 414–423.

Liberman, A. M., & Mattingly, I. G. (1989). A specialization for speech perception. *Science, 243,* 489–493.

Liberman, A. M., & Whalen, D. H. (2000). On the relation of speech to language. *Trends in Cognitive Sciences, 4,* 187–196.

Lima, C. F., & Castro, S. L. (2011). Speaking to the trained ear: Musical expertise enhances the recognition of emotions in speech prosody. *Emotion, 11,* 1021–1031.

Lindbergh, C. A. (1953). *The Spirit of St. Louis.* New York: Scribner.

London, J. (2004). *Hearing in time: Psychological aspects of musical meter.* New York: Oxford University Press.

Loui, P., Alsop, D., & Schlaug, G. (2009). Tone deafness: A new disconnection syndrome? *The Journal of Neuroscience, 29,* 10215–10220.

Loui, P., Li, H. C., Hohmann, A., & Schlaug, G. (2010). Enhanced cortical connectivity in absolute pitch musicians: A model for local hyperconnectivity. *Journal of Cognitive Neuroscience, 23,* 1015–1026.

Lurhmann, T. (2012). *When God talks back: Understanding the American evangelical relationship with God.* New York: Vintage.

Luria, A. R. (1969). *Traumatic aphasia.* The Hague: Mouton.

Luria, A. R., Tsvetkova, L. S., & Futer, D. S. (1965). Aphasia in a composer. *Journal of the Neurological Sciences, 2*, 288–292.

Machlis, J. (1977). *The enjoyment of music* (4th ed.). New York: Norton.

Majewski, W., Hollien, H., & Zalewski, J. (1972). Speaking fundamental frequency of Polish adult males. *Phonetica, 25*, 119–125.

Mampe, B., Friederici, A. D., Christophe, A., & Wermke, K. (2009). Newborns' cry melody is shaped by their native language. *Current Biology, 19*, 1994–1997.

Margulis, E. H. (2014). *On repeat: How music plays the mind*. New York: Oxford University Press.

Margulis, E. H., Simchy-Gross, R., & Black, J. L. (2015). Pronunciation difficulty, temporal regularity, and the speech to song illusion. *Frontiers in Psychology, 6*, 48.

Marie, C., Kujala, T., & Besson, M. (2012). Musical and linguistic expertise influence pre-attentive and attentive processing of non-speech sounds. *Cortex, 48*, 447–457.

Marques, C., Moreno, S., Castro, S., & Besson, M. (2007). Musicians detect pitch violation in a foreign language better than nonmusicians: Behavioral and electrophysiological evidence. *Journal of Cognitive Neuroscience, 19*, 1453–1463.

Mazzucchi, A., Marchini, C., Budai, R., & Parma, M. (1982). A case of receptive amusia with prominent timbre perception deficit. *Journal of Neurology, Neurosurgery and Psychiatry, 45*, 644–647.

McAdams, S. (2013). Musical timbre perception. In D. Deutsch (Ed.), *The psychology of music* (3rd ed., pp. 35–68). San Diego, CA: Elsevier.

McCarthy-Jones, S. (2012). *Hearing voices: The histories, causes and meanings of auditory verbal hallucinations*. New York: Cambridge University Press.

McGurk, H., & MacDonald, J. (1976). Hearing lips and seeing voices, *Nature, 264*, 746–748.

Meadows, J. C., & Munro, S. S. F. (1977). Palinopsia. *Journal of Neurosurgery and Psychiatry, 40*, 5–8.

Mebert, C. J., & Michel, G. F. (1980). Handedness in artists. In J. Herron (Ed.), *Neuropsychology of left-handedness* (pp. 273–280). New York: Academic Press

Mercer, T., & McKeown, D. (2010). Updating and feature overwriting in short-term memory for timbre. *Attention, Perception, & Psychophysics, 72*, 2289–2303.

Merkelbach, H., & van den Ven, V. (2001). Another white Christmas: Fantasy proneness and reports of "hallucinatory experiences" in undergraduate students. *Journal of Behavior Therapy and Experimental Psychiatry, 32*, 137–144.

Meyer, L. B. (1956). *Emotion and meaning in music*. Chicago, IL: University of Chicago Press.

Miller, G. A., & Heise, G. A. (1950). The trill threshold. *Journal of the Acoustical Society of America, 22*, 637–638.

Miller, G. A., & Licklider, J. C. R. (1950). The intelligibility of interrupted speech. *Journal of the Acoustical Society of America, 22*, 167–173.

Mithen, S. (2006). *The singing Neanderthals: The origins of music, language, mind, and body*. Cambridge, MA: Harvard University Press.

Miyazaki, K. (1990). The speed of musical pitch identification by absolute pitch possessors. *Music Perception, 8*, 177–188.

Miyazaki, K., Makomaska, S., & Rakowski, A. (2012). Prevalence of absolute pitch: A comparison between Japanese and Polish music students. *Journal of the Acoustical Society of America, 132*, 3484–3493.

Monsaingeon, B. (2001). *Sviatoslav Richter: Notebooks and conversations* (S. Spencer, Trans.). Princeton, NJ: Princeton University Press.

Moon, C., Cooper, R. P., & Fifer, W. P. (1993). Two-day-olds prefer their native language. *Infant Behavior and Development, 16*, 495–500.

Moreno, S., Marques, C., Santos, A., Santos, M., Castro, S. L., & Besson, M. (2009). Musical training influences linguistic abilities in 8-year-old children: More evidence for brain plasticity. *Cerebral Cortex, 19*, 712–723.

Mottron, L., Dawson, M., & Soulieres, I. (2009). Enhanced perception in savant syndrome: Patterns, structure and creativity. *Philosophical Transactions of the Royal Society of London, B., Biological Sciences, 364*, 1385–1391.

Mudd, S. A. (1963). Spatial stereotypes of four dimensions of pure tone. *Journal of Experimental Psychology, 66*, 347–352.

Nakajima, Y., Tsumura, T., Matsuura, S., Minami, H., & Teranishi, R. (1988). Dynamic pitch perception for complex tones derived from major triads. *Music Perception, 6*, 1–20.

Narmour, E. (1990). *The analysis and cognition of basic melodic structures: The implication-realization model*. Chicago: University of Chicago Press.

Necker. L. A. (1832). Observations on some remarkable phaenomena seen in Switzerland: and on an optical phaenomenon which occurs on viewing a crystal or geometrical solid. *The London and Edinburgh Philosophical Magazine and Journal of Science, 1*, 329–337.

Newport, E. L. (1990). Maturational constraints on language learning. *Cognitive Science, 14*, 11–28.

Norman-Haignere, S., Kanwisher, N. G., & McDermott, J. H. (2015). Distinct cortical pathways for music and speech revealed by hypothesis-free voxel decomposition. *Neuron, 88*, 1281–1296.

Nunes, J. C., Ordanini, A., & Valsesia, F. (2015). The power of repetition: Repetitive lyrics in a song increase processing fluency and drive market success, *Journal of Consumer Psychology, 25*, 187–199.

Oldfield, R. C. (1971). The assessment and analysis of handedness: The Edinburgh Inventory. *Neuropsychologia, 9*, 97–113.

O'Malley, F. W. (1920, October). Irving Berlin gives nine rules for writing popular songs. *American Magazine*.

Oxenham, A. J. (2013). The perception of musical tones, In D. Deutsch (Ed.), *The psychology of music* (3rd ed., pp. 1–34). San Diego, CA: Elsevier.

Pascual-Leone, A., Amedi, A., Fregni, F., & Merabet, L. B. (2005). The plastic human brain cortex. *Annual Review of Neuroscience, 28*, 377–401.

Patel, A. D. (2008). *Music, language, and the brain*. New York: Oxford University Press.

Patel, A. D., & Daniele, J. R. (2003). An empirical comparison of rhythm in language and music. *Cognition, 87*, B35–B45.

Penfield, W., & Perot, P. (1963). The brain's record of auditory and visual experience, *Brain, 86*, 595–696.

Penrose, L. S., & Penrose, R. (1958). Impossible objects: A special type of visual illusion. *British Journal of Psychology, 49*, 31–33.

Pereira, C. S., Teixeira, J., Figueiredo, P., Xavier, J., Castro, S. L., & Brattico, E. (2011). Music and emotions in the brain: Familiarity matters. *PLoS ONE, 6*(11), e27241.

Peretz, I., Ayotte, J., Zatorre, R. J., Mehler, J., Ahad, P., Penhune, V. B., & Jutras, B. (2002). Congenital amusia: A disorder of fine-grained pitch discrimination, *Neuron, 33*, 185–191.

Peretz, I., & Coltheart, M. (2003). Modularity of music processing. *Nature Neuroscience, 6*, 688–691.

Peretz, I., Cummings, S., & Dube, M.-P. (2007). The genetics of congenital amusia (tone deafness): A family-aggregation study. *American Journal of Human Genetics*, 81, 582–588.

Peretz, I., & Morais, J. (1989). Music and modularity. *Contemporary Music Review*, 1989, 4, 279–291.

Pfordresher, P. Q., & Brown, S. (2009). Enhanced production and perception of musical pitch in tone language speakers. *Attention, Perception, and Psychophysics*, 71, 1385–1398.

Piccirilli, M., Sciarma, T., & Luzzi, S. (2000). Modularity of music: Evidence from a case of pure amusia. *Journal of Neurology, Neurosurgery and Psychiatry*, 69, 541–545.

Pikler, A. G. (1966). Logarithmic frequency systems. *Journal of the Acoustical Society of America*, 39, 1102–1110.

Pinker, S. (1994). *The language instinct: How the mind creates language.* New York: Morrow.

Pinker, S. (1997). *How the mind works.* New York: Norton.

Polo, Marco (1958). *The Travels of Marco Polo.* (Trans. R. Latham.), New York: Penguin Books. (First published c. 1300).

Pousada, A. (1994). The multilinguism of Joseph Conrad. *English Studies*, 75, 335–349.

Profita, I. J., & Bidder, T. G. (1988). Perfect pitch. *American Journal of Medical Genetics*, 29, 763–771.

Querleu, D., Renard, X., Versyp. F., Paris-Delrue. L., & Crepin, G. (1988). Fetal hearing. *European Journal of Obstetrics and Gynecology and Reproductive Biology*, 28, 191–212.

Ragozzine, F., & Deutsch, D. (1994). A regional difference in perception of the tritone paradox within the United States. *Music Perception*, 12, 213–225.

Rauschecker, J. P., & Tian, B. (2000). Mechanisms and streams for processing of "what" and "where" in auditory cortex. *Proceedings of the National Academy of Sciences*, 97, 11800–11806.

Remez, R. E., Rubin, P. E., Pisoni, D. B., & Carell, T. D. (1981). Speech perception without traditional speech cues. *Science*, 212, 947–950.

Revesz, G. (1954). *Introduction to the psychology of music.* Norman, OK: University of Oklahoma Press.

Risset, J.-C. (1971). Paradoxes de hauteur: Le concept de hauteur sonore n'est pas le meme pour tout le monde. *Proceedings of the Seventh International Congress on Acoustics*, Budapest, S10, 613–616.

Rojas, D. C., Bawn, S. D., Benkers, T. L., Reite, M. L., & Rogers, S. J. (2002). Smaller left hemisphere planum temporale in adults with autistic disorder. *Neuroscience Letters*, 328, 237–240.

Rosenhan, D. L. (1973). On being sane in insane places. *Science*, 179, 250–258.

Rossing, T. D., Wheeler, T., & Moore, R. (2002). *The science of sound* (3rd ed.). San Francisco: Addison-Wesley.

Rubenstein, A. (1973). *My young years.* New York: Knopf.

Ruckmick, C. A. (1929). A new classification of tonal qualities. *Psychological Review*, 36, 172–180.

Sacks, O. (2008). *Musicophilia: Tales of music and the brain* (Rev. and Expanded). New York: Vintage.

Sacks, O. (2012). *Hallucinations.* New York: Knopf.

Saffran, J. R., & Griepentrog, G. (2001). Absolute pitch in infant auditory learning: Evidence for developmental reorganization. *Developmental Psychology*, 37, 74–85.

Sasaki, T. (1980). Sound restoration and temporal localization of noise in speech and music sounds. *Tohoku Psychologica Folia*, 39, 79–88.

Schellenberg, E. G. (2015). Music training and speech perception: A gene-environment interaction. *Annals of the New York Academy of Sciences*, 1337, 170–177.

Schellenberg, E. G., & Trehub, S. E. (2003). Good pitch memory is widespread. *Psychological Science, 14*, 262–266.

Schenker, H. (1956). *Neue Musikalische Theorien und Phantasien: Der Freie Satz.* Vienna, Austria: Universal Edition.

Schlaug, G., Jancke, L., Huang, Y., & Steinmetz, H. (1995). In vivo evidence of structural brain asymmetry in musicians. *Science, 267*, 699–701.

Semal, C., & Demany, L. (1991). Dissociation of pitch from timbre in auditory short-term memory. *Journal of the Acoustical Society of America, 89*, 2404–2410.

Sergeant, D. (1969). Experimental investigation of absolute pitch, *Journal of Research in Music Education, 17*, 135–143.

Shakespeare, W. *Antony and Cleopatra*, Act 4, Scene 14.

Shakespeare, W. *Julius Caesar*, Act 3, Scene 2.

Shakespeare, W. *The Tempest*, Act 4, Scene 1.

Shaw, G. B. (2007). *Pygmalion.* Minneapolis, MN: Filiquarian Publishing. (Original work published 1913)

Shepard, R. N. (1964). Circularity in judgments of relative pitch. *Journal of the Acoustical Society of America, 36*, 2345–2353.

Shepard, R. N. (1965). Approximation to uniform gradients of generalization by monotone transformations of scale. In D. I. Mostofsky (Ed.), *Stimulus generalization* (pp. 94–110). Stanford, CA: Stanford University Press.

Shepard, R. N. (1982). Structural representations of musical pitch. In D. Deutsch (Ed.), *The psychology of music* (pp. 343–390): New York: Academic Press.

Shermer, M. (2009, January 1). Telephone to the dead. *Scientific American.*

Shermer, M. (2011). *The believing brain.* New York: St. Martin's Press.

Sidgwick, H., Johnson, A., Myers, F. W. H., Podmore, F., & Sidgwick, E. M. (1894). Report on the census of hallucinations. *Proceedings of the Society for Psychical Research, 10*, 24–422.

Simon, H. A. (1962). The architecture of complexity. *Proceedings of the American Philosophical Society, 106*, 467–482.

Simpson, J. (1988). *Touching the void: The true story of one man's miraculous survival.* New York: HarperCollins.

Sloboda, J. A. (1985). *The musical mind: The cognitive psychology of music.* Oxford, UK: Oxford University Press.

Smith, N. A., & Schmuckler, M. A. (2008). Dial A440 for absolute pitch: Absolute pitch memory by nonabsolute pitch possessors. *The Journal of the Acoustical Society of America, 123*, EL77–EL84.

Spence, M. J., & Freeman, M. S. (1996). Newborn infants prefer the maternal low-pass filtered voice, but not the maternal whispered voice. *Infant Behavior and Development, 19*, 199–212.

Spencer, H. (1857, October). The origin and function of music. *Fraser's Magazine.*

Steele, J. (1923). *Prosodia rationalis: Or, an essay towards establishing the melody and measure of speech to be expressed and perpetuated by peculiar symbols* (2nd ed, amended and enlarged). London: J. Nicols. (Original work published 1779)

Strauss, R. (1953). *Recollections and reflections.* (W. Schuh, Ed, E. J. Lawrence, Trans.). Zurich, Switzerland: Atlantis-Verlag. (Original work published 1949)

Suedfeld, P., & Mocellin, J. S. P. (1987). The "sensed presence" in unusual environments. *Environment and Behavior, 19*, 33–52.

Starr, G. E., & Pitt, M. A. (1997). Interference effects in short-term memory for timbre. *Journal of the Acoustical Society of America, 102*, 486–494.

Sundberg, J. (2013). Perception of singing. In D. Deutsch (Ed.), *The psychology of music* (3rd ed., pp. 69–106). San Diego, CA: Elsevier.

Takeuchi, A. H., & Hulse, S. H. (1991). Absolute pitch judgments of black and white-key pitches. *Music Perception, 9*, 27–46.

Temperley, D. (2007). *Music and probability.* Cambridge, MA: MIT Press.

Teranishi, R. (1982). Endlessly ascending/descending chords performable on a piano. *Reports of the Acoustical Society of Japan*, H62–H68.

Terhardt, E., & Seewann, M. (1983). Aural key identification and its relationship to absolute pitch, *Music Perception, 1*, 63–83.

Thiessen, E. D., Hill, E. A., & Saffran, J. R. (2005). Infant-directed speech facilitates word segmentation. *Infancy, 7*, 53–71.

Thompson, W. F. (2013). Intervals and scales. In D. Deutsch (Ed.), *The psychology of music* (3rd ed., pp. 107–140), San Diego, CA: Elsevier.

Thompson, W. F., Hall, M. D., & Pressing, J. (2001). Illusory conjunctions of pitch and duration in unfamiliar tonal sequences. *Journal of Experimental Psychology: Human Perception and Performance, 27*, 128–140.

Thompson, W. F., Marin, M. M., & Stewart, L. (2012), Reduced sensitivity to emotional prosody in congenital amusia rekindles the musical protolanguage hypothesis. *Proceedings of the National Academy of Sciences, 109*, 19027–19032.

Thompson, W. F., Schellenberg, E. G., & Husain, G. (2004). Decoding speech prosody: Do music lessons help? *Emotion, 4*, 46–64.

Thomson, W. (1999). *Tonality in music: A general theory.* San Marino, CA: Everett.

Tien, A. Y. (1991). Distribution of hallucinations in the population. *Social Psychiatry and Psychiatric Epidemiology, 26*, 287–292.

Tierney, A., Dick, F., Deutsch, D., & Sereno, M. (2013). Speech versus song: Multiple pitch-sensitive areas revealed by a naturally occurring musical illusion. *Cerebral Cortex, 23*, 249–254.

Treisman, A. M., & Gelade, G. (1980). A feature-integration theory of attention. *Cognitive Psychology, 12*, 97–136.

Twain, M. (1876). A literary nightmare. *Atlantic Monthly, 37*, 167–170.

Van Hedger, S. C., Heald, S. L. M., & Nusbaum, H. C. (2016). What the [bleep]? Enhanced pitch memory for a 1000 Hz sine tone. *Cognition, 154*, 139–150.

Van Noorden, L. P. A. S. (1975). *Temporal coherence in the perception of tone sequences* (Unpublished doctoral dissertation). Technische Hogeschoel Eindhoven, the Netherlands.

Vanden Bosch der Nederlanden, C. V. M., Hannon, E. E., & Snyder, J. S. (2015a). Everyday musical experience is sufficient to perceive the speech-to-song illusion. *Journal of Experimental Psychology: General, 144*, e43–e49.

Vanden Bosch der Nederlanden, C. V. M., Hannon, E. E., & Snyder, J. S. (2015b). Finding the music of speech: Musical knowledge influences pitch processing in speech. *Cognition, 143*, 135–140.

Varney, N. R., & Benton, A. L. (1975). Tactile perception of direction in relation to handedness and familial handedness. *Neuropsychologia, 13*, 449–454.

Vargha-Khadem, F., Carr, L. J., Isaacs, E., Brett, E., Adams, C., & Mishkin, M. (1997). Onset of speech after left hemispherectomy in a nine year old boy. *Brain, 120*, 159–182.

Vernon, P. E. Absolute pitch: A case study. (1977). *British Journal of Psychology, 68*, 485–489.

Vokey, J. R., & Read, J. D. (1985). Subliminal messages: Between the devil and the media. *American Psychologist, 40*, 1231–1239.

Wallach, H., Newman, E. B., & Rosenzweig, M. R. (1949). The precedence effect in sound local-ization. *The American Journal of Psychology, 62,* 315–336.

Warren, R. M. (1983). Auditory illusions and their relation to mechanisms normally enhancing accuracy of perception. *Journal of the Audio Engineering Society, 31,* 623–629.

Warren, R. M., Obusek, C. J., & Ackroff, J. M. (1972). Auditory induction: Perceptual synthesis of absent sounds. *Science, 176,* 1149–1151.

Warren, R. M., Obusek, C. J., Farmer, R. M., & Warren, R. P. (1969). Auditory se-quence: Confusions of patterns other than speech or music. *Science, 164,* 586–587.

Wessel, D. L. (1979). Timbre space as a musical control structure. *Computer Music Journal, 3,* 45–52.

White, R. W. (1964). Criminal complaints: A true account by L. Percy King. In B. Kaplan (Ed.), *The inner world of mental illness* (pp. 134–135). New York: Harper.

Williamson, V. J., Jilka, S. R., Fry, J., Finkel, S., Mullensiefen, D., & Stewart, L. (2012). How do "earworms" start? Classifying the everyday circumstances of involuntary musical imagery. *Psychology of Music, 40,* 259–284.

Winkielman, P., Schwarz, N., Fazendeiro, T., & Reber, R. (2003). The hedonic marking of pro-cessing fluency: Implications for evaluative judgment. In J. Musch & K. C. Klauer (Eds.), *The psychology of evaluation: Affective processes in cognition and emotion* (pp. 189–217). Mahwah, NJ: Erlbaum.

Woods, B. T. (1983). Is the left hemisphere specialized for language at birth? *Trends in Neurosciences, 6,* 115–117.

Wong, P. C., Skoe, E., Russo, N. M., Dees, T., & Kraus, N. (2007). Musical experience shapes human brainstem encoding of linguistic pitch patterns. *Nature Neuroscience, 10,* 420–422.

Wray, A. (1998). Protolanguage as a holistic system for social interaction. *Language and Communication, 18,* 47–67.

Wright, D. (1969). *Deafness: A personal account.* New York: HarperCollins.

Yam, P. (1996, March). Escher for the ear. *Scientific American.*

Zajonc, R. B. (1968) Attitudinal effects of mere exposure. *Journal of Personality and Social Psychology, 9,* 1–27.

Zatorre, R. J., Belin, P., & Penhune, V. B. (2002). Structure and function of auditory cortex: Music and speech. *Trends in Cognitive Science, 6,* 37–46.

# INDEX

Page numbers in bold indicate illustrations.